Native & Naturalized
WOODY PLANTS
of Austin & the Hill Country

Brother Daniel Lynch, C.S.C.

Illustrated by Nancy McGowan

Sponsored by

Travis Audubon Society

Saint Edward's University

Second Printing 1994

ISBN 0-938472-00-3
Library of Congress Catalog Card Number 80-53737

Lynch, Daniel
   Native and naturalized woody plants of Austin and the hill country.

Austin, Texas: Saint Edward's Univ.

180 p.

8011   800924

Edited by *Jane Moseley*
Designed by *Barbara Whitehead*
Printed by *Acorn Press*

CONTENTS

INTRODUCTION

T HIS book is an introduction to 142 woody, or seemingly woody, plants that grow in a wild state in and around Austin and into the Hill Country. Because these plants have a general northeast to southwest distribution on either side of Interstate 35, the book is useful from Georgetown and somewhat beyond, southwest to New Braunfels. Included are the woody plants that follow stream courses eastward into the Blackland Prairie and those that grow westward on the limestone of the Edwards Plateau to the Central Mineral Region. (See map on page viii .)

The book was written primarily for those who perhaps can identify a few trees, shrubs, and woody vines but would like to identify more of them. Nancy McGowans's accurate and detailed drawings from live specimens give the angle, the tilt, and the bend characteristic of twig and leaf as seen in the field. The accompanying botanical key and plant descriptions provide an easy and direct means of identification.

Technical terms have been avoided as much as possible. Those that are used are explained in the illustrated glossary. The key to identification relies heavily on leaves, structures readily visible to the observer and present on the plant most of the year. Leaves show enough variety in arrangement, size, shape, texture, and color to provide a reliable means of distinguishing one species from another. Identification by leaf structure is supported by description of bark, flowers, fruits, and habitat.

Not all the distinguishing features of the woody plants are visible to the naked eye. Some need to be magnified. Others that are visible need to be measured. It is recommended that a 10x hand lens and a 6-inch ruler be taken on walks or field trips to aid in identification.

Plants described in the book are grouped according to families. Both colloquial and Latin names are given. In most cases, there are more colloquial names in the literature than I think should be included in a book of local coverage. An effort was made to select the most commonly used ones. Frequency of occurence in references plus the opinions of persons familiar with the Central Texas flora were the main criteria for selection. The selected references give additional names. The Latin, or scientific name, is a *binomial*, a two-part name giving genus and species. (Example: The oaks are in the genus *Quercus*. The Bur Oak, which has a large acorn, is *Quercus macrocarpa*. The Texas Oak is *Quercus texana*. When the specific epithet is taken from a person's name, e.g. *Quercus Shumardii*, Shumard Oak, the species is often capitalized.) Only one scientific name is given, and, unlike colloquial names, only one is technically correct. Unfortunately, many Texas plants were described and named when communication among botanists in North America and Europe was difficult. Botanists studying the same flora, but isolated from one another, would describe the same plants under different names. Also, today, as more information about native plants comes to light, it is sometimes necessary to change scientific names.

Many references list plants under what is regarded as their correct or preferred scientific name and add synonyms. The nomenclature of this book, with only two exceptions, follows Correll and Johnston (1970). Colloquial names are mostly from Correll and Johnston, Gould (1975), and Vines (1960). In a few cases, variety, or subspecies, follows the specific epithet. This designates a particular section of the species. The author of a scientific name is indicated by placing that person's name, usually in abbreviated form, after the plant name. An abbreviated name in parentheses, e.g. (L.), indicates a prior author Linnaeus.

The distinction between *woody* and *herbaceous* when part of the plant overwinters above ground and contains hard tissue is not always a sharp one. Following a strict definition of *woody* would eliminate from this book some plants of this type which form an important part of our flora. Because they are seemingly *woody* and are perennial they are included.

This is a guide to native and *naturalized* plants. Native plants are indigenous to the region in which they are found. *Naturalized* plants have been introduced from another region, often as *garden escapes*, and are now reproducing themselves in the wild and becoming part of the regional flora. Plants that have escaped cultivation and have become established by seed in large enough numbers to assure them a permanent place in the flora are included. Escapes that have not spread beyond the vicinity of homes or have not established what would appear to be viable populations in the wild are omitted.

Native and naturalized plants are obviously adapted to the habitats in which they are growing and usually require little care when included in a landscape design. Many are attractive and beautify both the urban and wild landscape. Why remove them? Why not design around them, leaving them in their natural state? A word of caution. Wild trees, shrubs, and woody vines, as a rule, have particular habitat requirements and often fail to survive transplanting. A number of native and naturalized woody plants are raised in nurseries and can be purchased. These have a much better chance of survival than those dug up in the wild. Rare plants should never be transplanted unless it is to save them from destruction.

In conservation efforts, it is well to remember that we cannot appreciate what we do not know. Perhaps this book will lead to greater knowledge of native and naturalized trees, shrubs, and vines around homes, neighborhoods, and countryside—thereby bringing about appreciation and conservation of them.

D.L. Austin, Texas

ACKNOWLEDGEMENTS

I am grateful to Dr. Marshall C. Johnston of the Department of Botany of The University of Texas and to Mr. Edward A. Kutac for assistance in locating specimen plants, and to Mr. and Mrs. Edward M. Norris, Mrs. Jewel McDonald and the late Mr. Robert E. McDonald, the late Mr. and Mrs. Smith W. Ligon, Travis Ecology, and Texas System of Natural Laboratories for access to natural areas which furnished essential habitat information.

I thank many for their interest and support: especially members of the Now or Never group and the Travis Audubon Society, and Mr. David Mahler for furnishing the names of plants in the Wild Basin Wilderness Park, Mr. John Scanlan, Attorney, for his assistance, and Mrs. Jane Moseley for her contribution as coordinator of the project and editor of the book.

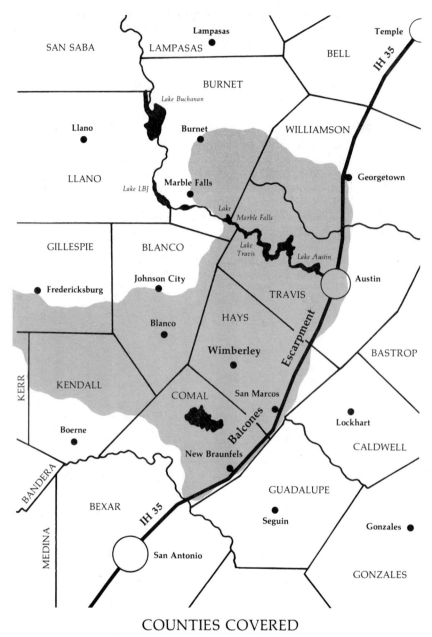

COUNTIES COVERED

Coverage can be expected to decrease with distance from shaded area.

Many of the plants in this book can be found in the Wild Basin Wilderness Park on Bee Creek, just west of Austin. Such plants are designated with an asterisk.*

Native and Naturalized
WOODY PLANTS

ix

x

xi

ILLUSTRATIONS

Bald Cypress
Taxodium distichum (L.) Rich.

Grows to a majestic size, the largest tree in Austin. Found at the edges of Lake Austin and Town Lake and along the banks of streams. Branchlets with feathery leaves. Leaves ½ to ¾ inch long, soft, turning yellow in autumn. A cone-bearing tree with round female cones 1 inch in diameter and pendulous clusters of small male cones.

Ashe Juniper, Post Cedar, Mountain Cedar*
Juniperus Ashei Buchh.

An evergreen tree with branches commonly arising from the base
of the trunk. Abundant on limestone hills. Leaves closely
appressed, minute and scalelike. Female trees with blue, berrylike
cones; male with a burnt gold appearance in winter due to pollen.

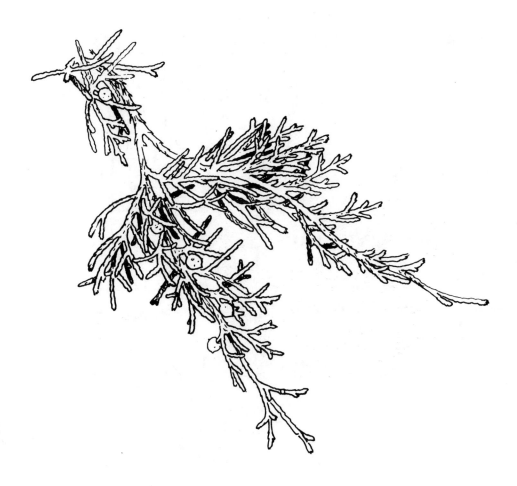

Eastern Red Cedar
Juniperus virginiana L.

Similar to Ashe Juniper but without branching from the base and with a narrower crown. Much less common than Ashe Juniper, but the common juniper in and around Bastrop.

Joint-fur
Ephedra antisyphilitica C.A. Mey.

An upright or spreading shrub found on limestone hillsides, on bluffs, and in gravelly or stony upland soils. Jointed branches green to yellowish or grayish green, with a light orange to yellow or tan band encircling the stem below the leaves. Leaves minute, scalelike at the joints. Reproductive structures in small cones, ¼ to ½ inch long, emerging from the joints. Not to be confused with either Horsetail or Scouring-rush, genus *Equisetum*, which are herbaceous. Less frequent in the Austin area than west of it.

Giant Reed
Arundo Donax L.

A tall perennial grass growing in large clumps. Naturalized from the Old World, frequent in ditches and in low, open places. Stems unbranched, up to 16 feet tall. Leaves (blades) narrow, up to 2 feet in length. Flower parts, other than the pistil and stamens, chaffy. Flowers in dense, elongate clusters topping the stems.

Dwarf Palmetto, Bush Palmetto
Sabal minor (Jacq.) Pers.

A palm of low, wet places; at the western limit of its range in the eastern part of the Hill Country. Usually stemless, the leaves arising from an underground stock. Leaf blades longer than the leaf stalks, fan shaped, as much as 4 feet wide, dissected, the narrow segments notched at the tip. Flowers small, white, numerous in showy clusters. Fruit spherical, about ½ inch wide, black.

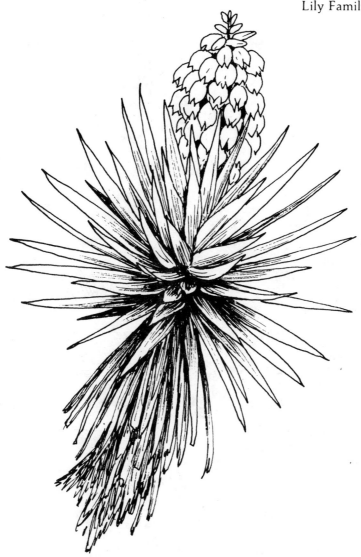

Spanish Dagger, Trecul Yucca
Yucca Treculeana Carr.

Shrub or small tree with stem unbranched except near the top, and up to 10 or more feet tall. In brushy and open areas west of Austin. Leaves stiff, sharp pointed, up to 3 feet long by 3 inches wide, in large clumps at the ends of stems or branches, dead ones hanging below the live ones. Flowers white or faintly purplish, in dense, showy clusters rising above the leaves, appearing in March and April. Fruit a capsule up to 4 inches long by 1 inch wide.

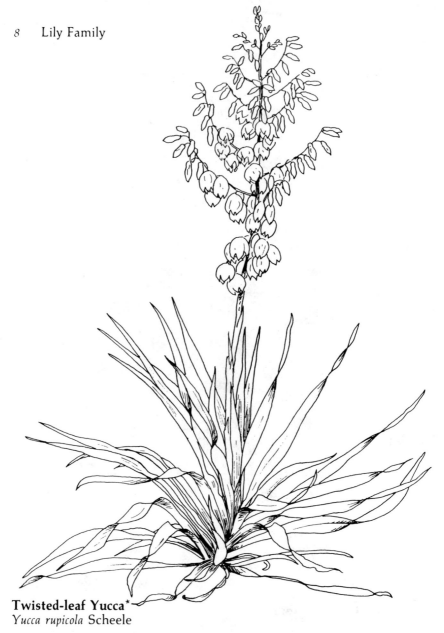

Twisted-leaf Yucca[*]
Yucca rupicola Scheele

The common yucca of the Austin area. Similar to *Yucca arkansana* but larger. Leaves fewer than in *Y. arkansana*, twisted, and up to 2 feet long, without curly fibers on their margins. Flowering stalks often over 5 feet tall, bearing a cluster of bell-shaped, white flowers with petals up to 2½ inches long and an inch wide, appearing from April to June. On shallow, stony soil in fields, among scattered trees, and on canyon ledges.

Buckley Yucca
Yucca constricta Buckl.

Occasional in brushy and open areas. Stem below the leaves often
not present; when present more or less prostrate. Leaves stiff,
sharp pointed, up to 2 feet long by ⅝ inch wide, with white fibers
curling from their margins, clustered at the ends of stems or at the
ground line. Flowers showy, greenish white, bell shaped, up to 2
inches long by 1 inch wide, borne in a large, branched cluster on a
tall stalk, the stalk and flowers combined as much as 10 feet tall.
Fruit a capsule up to 2½ inches long by 1½ inches wide, opening
from the tip.

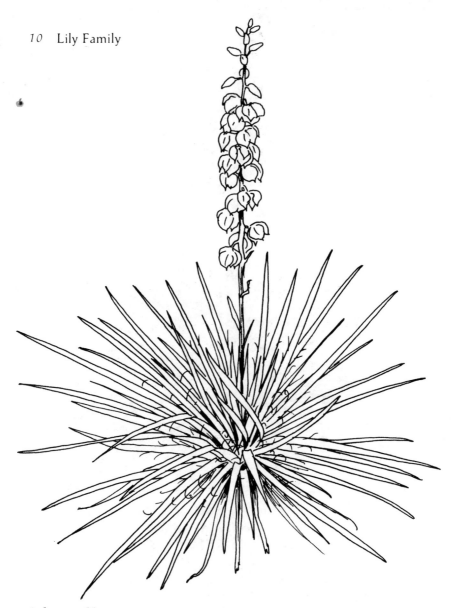

Arkansas Yucca
Yucca arkansana Trel.

In open fields and at the edges of thickets on upland soils. Leaves from the base, sharp pointed, with curly white fibers on the margins. Flower stalk erect, rising from the clump of basal leaves, usually unbranched. Flowers with 3 showy, greenish white, petals and 3 similar sepals, appearing from April to June. Fruit a capsule, opening from the tip when dry. Seeds many, flat, waferlike, black when mature.

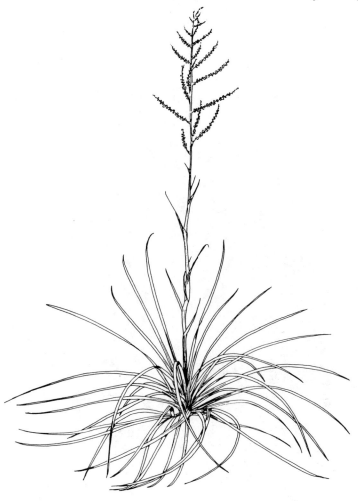

Devil's-shoestring, Ribbon Grass*
Nolina Lindheimeriana (Scheele) Wats.

Resembling a large clump of grass in the open and in light shade on limestone slopes and cliffs. Stem usually single, herbaceous, from a woody underground base. Leaves arising from the ground line or from the lower part of the stem, long, flat, and narrow, up to 30 inches long by 5/16 inch wide, with finely serrated margins, teeth visible under a 10x hand lens. Flowers about 3/16 inch wide, white to cream, numerous on slender branches from the upper part of the rather tall stem and forming a loose, open display 1 to 3 feet or more tall and several inches wide, opening in spring and early summer. Fruit a short capsule about 3/16 inch wide.

Sacahuista, Bunch-grass*
Nolina texana Wats.

Resembling a large clump of coarse grass. Conspicuous on wooded limestone ledges and slopes. Stem herbaceous, often more than one, arising from a woody base. Leaves evergreen, firm textured, long, thin, up to 4 feet long by 3/16 inch wide, forming a thick clump. Flowers small, white to cream, in dense vertical clusters 10 inches or more long by 2 inches or more wide at the tops of rather short stems, appearing in spring and early summer. Fruit distinctly 3-lobed, a thin-walled capsule 3/16 to ¼ inch wide.

Sotol
Dasylirion texanum Scheele

On open, rocky sites west of Austin; at the falls in Pedernales **Falls** State Park. Stem supporting a dense basal clump of long, narrow leaves. Leaf blades to 3 feet or more in length and ½ inch in width, with a frayed tip, spiny margins, and a broad, smooth-margined base. Flowers minute, in a conspicuous, dense, spikelike cluster up to 3 feet long atop a 5 to 12 foot stalk. Fruit a small, 3-winged capsule.

Cat-brier, Saw Greenbriar[*]
Smilax Bona-nox L.

A prickly vine forming dense tangles in shrubby and wooded areas. Stems smooth, green, with stout, sharp prickles on the lower sections. Leaves with tendril-bearing petioles; blades up to 4½ inches long and 4 inches wide, varying from triangular to heart shaped, often with a broad lobe on each side, firm textured and occasionally mottled on the upper surface, persisting into winter. Flowers small, rather inconspicuous, in clusters arising from the axils of the leaves, male and female on different plants, appearing from March to June. Fruit spherical, ¼ inch in diameter, fleshy, black.

Black Willow*
Salix nigra Marsh.

A fast-growing tree, often with several trunks growing out at angles from one root. Found in wet soil along streams and at the margins of ponds and lakes. Leaf blades up to 5 inches long, narrow and tapering to an elongate tip, margins finely serrate. Flowers inconspicuous, arranged in elongate clusters which appear in March and April; male and female flowers on separate trees. Seeds wind-borne on silky hairs. Two varieties in the Austin area: var. *nigra* and var. *Lindheimeri*, the latter with narrower leaves tapering more gradually to the base.

White Poplar, Alamo Blanco
Populus alba L.

Tree planted around homes and escaping to roadsides and
fencerows, forming small stands through rootsprouts. Native of
Europe and Asia. With gray to white bark becoming roughened
with age; leaf-bearing shoots with soft, white hairs. Leaf blades
about as broad as long to broader than long, usually with 2
unequal lobes on each side and a terminal lobe. Blades with
toothed margins, more or less flat across the base, the upper
surface green, the lower surface conspicuously white, covered with
a dense mat of wooly hairs. Flowers without petals, in pendulous
clusters 2 to 3 inches long, male and female on separate trees,
appearing before the leaves. Fruit a small capsule. Seeds
windblown on a tuft of white hairs.

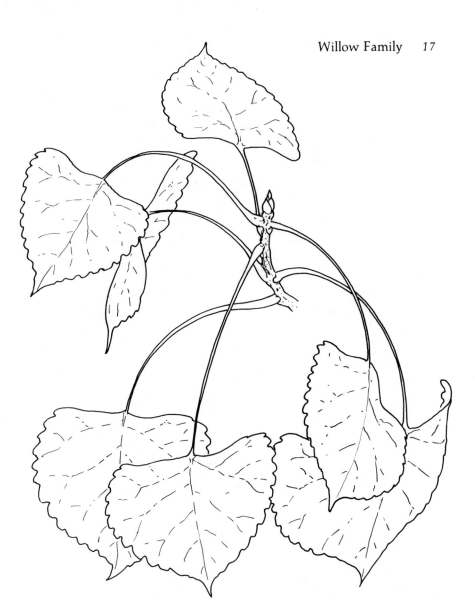

Eastern Cottonwood, Alamo
Populus deltoides Marsh.

A fast-growing tree that sometimes attains a large size. The common cottonwood along waterways and in wet ditches. Leaf blades triangular, with a smooth surface, scalloped margins, and an elongate tip. Pendulous clusters of flowers without petals in late March and early April. Seeds wind-borne on a tuft of cottony hairs.

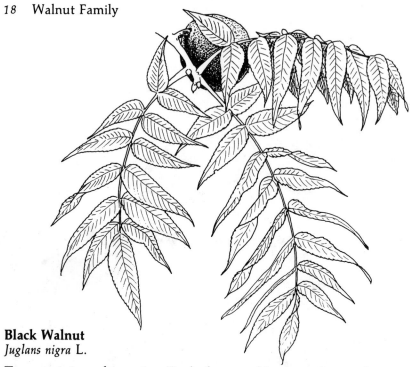

Black Walnut
Juglans nigra L.

Tree attaining a large size. Dark, furrowed bark on the trunk.
Leaves up to 2 feet long with 5 to 11 pairs of leaflets along a
central axis and a single leaflet at the tip; midrib of the lateral
leaflets off-center with the wider part of the blade toward the leaf
tip. Flowers inconspicuous, in elongate, green clusters. Fruit 1½ to
2¼ inches in diameter, consisting of a hard-shelled, furrowed nut
enclosed in a green husk, darker when ripe. Found mostly east of
the Hill Country, occasionally planted. A valuable hardwood tree,
prized for furniture and paneling.

Arizona Walnut *
Juglans major (Torr.) Heller

Not Illustrated

A broad-crowned tree found in canyons and along streams in the
Hill Country. With furrowed bark on the trunk, gray twigs, and
reddish brown new growth covered with hairs visible under a 10x
hand lens. Leaves up to 1 foot long with 4 to 8 pairs of leaflets
along a central axis and a single leaflet at the tip, the margins
finely toothed and the midrib of the lateral leaflets off-center, the
wider part of the blade toward the leaf tip. Flowers inconspicuous,
in elongate green clusters. Fruit spherical, 1 to 1½ inches in
diameter, intermediate in size between Black Walnut and Little
Walnut.

Little Walnut, River Walnut *
Juglans microcarpa Berl.

Small tree, often shrubby with several trunks. Found in canyons and along streams. Bark furrowed to smooth on trunks, gray on older twigs, and reddish on new growth. Leaves and fruits smaller than those of the Black Walnut, husk not more than 1 inch in diameter.

Pecan
Carya illinoiensis (Wang.) K. Koch

The state tree of Texas. Grows to a large size, found in bottomlands. Now represented by a number of cultivated varieties. Leaves up to 1½ feet long with 4 to 8 pairs of leaflets from a central axis and a single leaflet at the tip. Midrib of the leaflet off-center with the wider part of the blade toward the leaf tip. Flowers inconspicuous, male in elongate clusters, both sexes on same tree. Fruit a nut enclosed in a thin husk splitting open at maturity, husk often persistent on the tree for weeks after the nut has fallen.

Bur Oak
Quercus macrocarpa Michx.

Tree occasionally reaching a large size. Found on stream terraces and floodplains. Leaves up to 9 inches long with a central midrib from which branch veins lead into rounded lobes. Lobes separated by deep sinuses reaching, in some cases, to within ½ inch of the midrib. Lobes beyond the midpoint of the blade wavy margined and longer and broader than those toward the base. Acorns large, up to 1½ inches broad with ¼ to more than ½ of the acorn enclosed in the cup. Cup with coarse scales and a fringed margin.

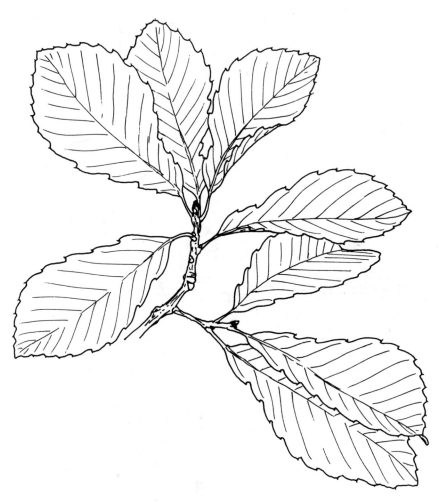

Chinkapin Oak, Chinquapin Oak
Quercus Muehlenbergii Engelm.

A tree with light gray platy or scaly bark and smooth, gray twigs changing to brown on the current year's leaf-bearing growth. On limestone slopes in the Hill Country. Leaves up to 8 inches long and 4½ inches wide with their widest part nearer the apex than the base. Larger leaves broadly rounded from the widest part to the apex and tapered to the base, the smaller ones narrower; leaf margins shallowly lobed or coarsely toothed, each lobe or tooth with a minute tip; the upper surface smooth, with a sheen, the lower surface dull. Flowers inconspicuous in narrow clusters. Fruit an acorn up to 1 inch long and ¾ inch wide.

Post Oak
Quercus stellata Wang. var. stellata

A medium-sized tree found on gravelly soils. Associated with
Blackjack Oak in Bastrop State Park. Leaf blades variable, 3 to 5
inches long or longer, wavy margined to deeply lobed; the lobes
rounded and up to 4 on each side, the upper pair often much
larger than the others. Acorns up to ¾ inch long, sometimes to 1¼
inches, the cup without the fringe found in Bur Oak.

White Shin Oak *

Quercus sinuata var. *breviloba* (Torr.) C.H. Mull.

A shrub or small tree with light gray, flaking bark. Forms thickets on shallow soil, seldom a single tree, found mostly on flat-topped limestone hills. Leaves up to 3 inches long, irregularly and shallowly lobed with their broadest part nearer the tip than the base. Acorns up to ⅝ inch long and ⅜ inch wide.

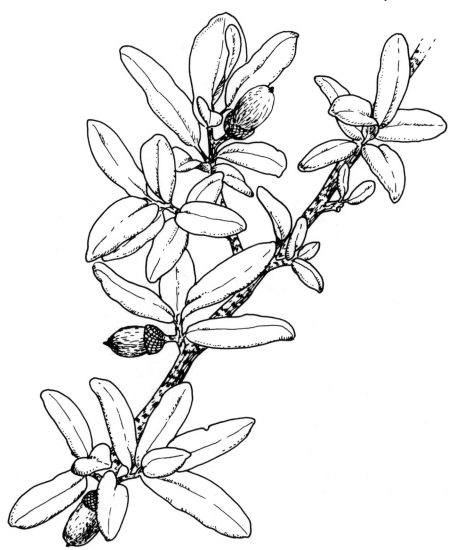

Plateau Live Oak *

Quercus fusiformis Small

The large, spreading Live Oak of Austin neighborhoods. Grows as a tree or shrub throughout the Hill Country. Leaves evergreen, firm textured, ovate to elliptic, 1 to 3 inches long; usually without lobes except on young plants and rootsprouts, then with pointed lobes. Acorns ¾ to 1 inch long, rather elongate.

Shumard Oak, Southern Red Oak *
Quercus Shumardii Buckl.

An oak of canyons and narrow valleys. Leaves frequently turning scarlet in the fall, up to 7 inches long, with 2 to 4 pairs of pointed lobes with soft, bristlelike tips. Sinuses between the lobes reaching from ½ to ¾ the distance from the tip of the lobe to the leaf midrib. Acorns almost as wide as long, ¾ to 1 inch long when mature with a broadly rounded apex and a flat base.

Texas Oak, Spanish Oak *
Quercus buckleyi Nixon and Dorr

Commonly associated with Plateau Live Oak and Ashe Juniper on limestone hills. Similar to Shumard Oak but leaves smaller, up to 5½ inches long, with narrower lobes and deeper sinuses, the latter sometimes reaching to ¼ inch of the midrib. Acorns egg shaped and up to ¾ inch long.

Blackjack Oak
Quercus marilandica Muenchh.

A medium-sized tree of gravelly soils. Abundant in Bastrop State Park. Leaf blades wedge shaped with a narrow, rounded base and broadening toward the tip. Blades shallowly lobed with usually an apical and 2 lateral lobes bearing bristlelike extensions of the main vein. Acorn elliptic, broadly rounded at the apex and base, up to ¾ inch long when mature.

Sugar Hackberry, Texas Sugarberry, Palo Blanco *
Celtis laevigata Willd.

Often comes up in vacant lots and in fencerows, common along
streams and in bottomlands. Frequently heavily parasitized by
mistletoe. Bark grayish, warty on older trunks and limbs. Leaves
up to 4 inches long, blades ovate to narrower with a long, tapering
tip, usually with smooth margins and an unequal base which is
tapered on one side of the midrib and rounded on the other. Fruit
spherical, ¼ inch in diameter and usually dull red.

Netleaf Hackberry, Palo Blanco *

Celtis reticulata Torr.

A tree or large shrub with gray bark, ridged on the trunk, smooth on the twigs. Found on wooded, limestone slopes. Leaves up to 3 inches long, smaller than those of Sugar Hackberry, with smooth margins, sometimes with teeth in the apical half, a pointed tip, and a slightly asymmetric base, the upper surface somewhat rough to the touch and darker green than the lower surface, midrib and veins light yellow on the lower. Flowers inconspicuous. Fruit spherical, ¼ inch in diameter, reddish.

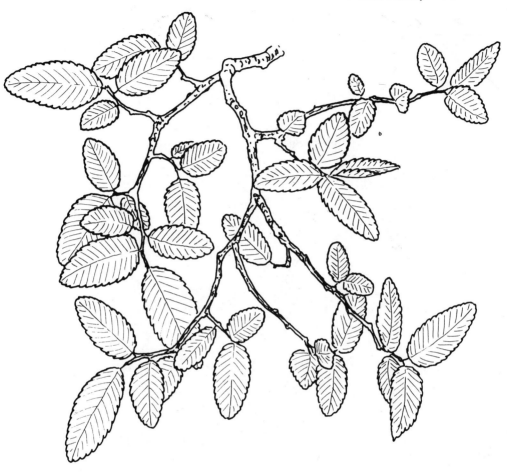

Cedar Elm, Olmo *
Ulmus crassifolia Nutt.

A common tree of wooded areas on limestone. Branchlets of young
trees with corky wings. Leaves much smaller than those of the
American Elm, generally not more than 2½ inches long. Flowers
and fruit similar to those of the American Elm but appearing in
the fall.

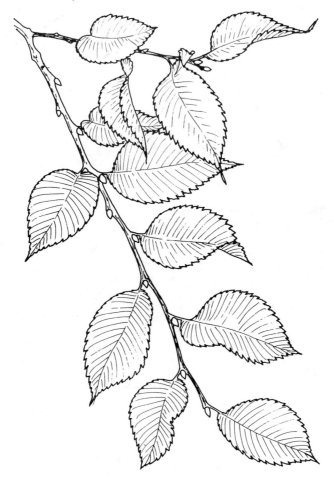

American Elm, White Elm [*]
Ulmus americana L.

A popular shade tree with a vaselike shape, fine specimens on the
Capitol grounds; native to stream banks and bottomlands. Leaves
ovate, up to 6 inches long with an elongate tip, serrate margins,
and an unequal base which is rounded on one side of the blade and
tapered on the other. Flowers inconspicuous, appearing in
February before the leaves, followed in March by waferlike fruits
⅜ inch long by ¼ inch wide. Severely reduced in numbers in the
Northeast and Midwest by Dutch Elm disease but not infected in
the Austin and Hill Country area.

Texas Mulberry, Mountain Mulberry *
Morus microphylla Buckl.

Shrub or small tree with smooth, light gray bark. In canyons and
on limestone slopes, rather infrequent in the Austin area. Leaves
smaller than those of red and white mulberry; blades up to 2½
inches long, roughly ovate, frequently lobed, with toothed
margins, an extended tip, and rounded or slightly lobed base.
Flowers inconspicuous, in short, drooping clusters. Fruit a cluster
of minute, fleshy, berrylike fruits varying from red to black,
ripening in May, edible.

White Mulberry, Silkworm Mulberry, Moral Blanco
Morus alba L.

A small tree naturalized from China and found near streams and in woodlands. Forms of this species provide leaves fed to silkworms. Leaves ovate and up to 8 inches long, often irregularly lobed with lobes on one or both sides of the blade and the sinuses between them deep or shallow. Leaf tip narrow and pointed or blunt, margins serrate, and the base of the blade rounded or lobed. Flowers inconspicuous, in short spikes. Fruit fleshy, resembling an elongate cluster of minute berries, white to red or purple, edible. Fruit ½ to ¾ inch long.

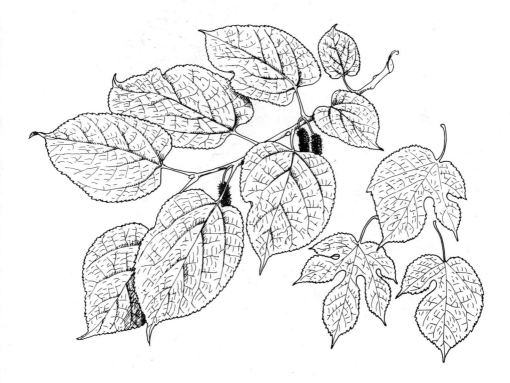

Red Mulberry, Moral
Morus rubra L.

Habitat similar to that of the White Mulberry. Leaves up to 8
inches long, ovate and with or without lobes; tip elongate,
emerging abruptly from the rounded blade. Blade margins serrate
and base rounded or truncate to somewhat heart shaped, lower
surface covered with fine hairs and soft to the touch, turning
bright yellow in the fall. Flowers and fruits similar to those of the
White Mulberry. Fruit ¾ to 1¼ inches long.

Bois-d'arc, Osage Orange, Horse Apple, Naranjo Chino
Maclura pomifera (Raf.) Schneid.

A durable tree, once planted in hedgerows; furnished bow wood
for the Osage Indians. Bark yellowish brown, furrowed. Small
branches with thorns up to 1 inch long. Leaves, including petiole,
up to 9 inches long, shiny, ovate to narrower with a smooth
margin. Flowers inconspicuous. Fruit conspicuous, green, the size
and shape of an orange or grapefruit and containing a milky sap,
inedible. Male and female flowers on separate trees, so fruits not
on all trees.

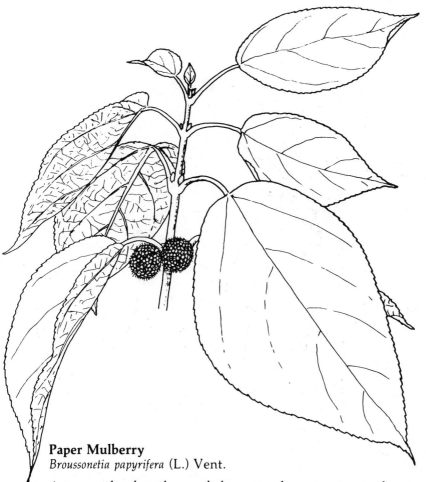

Paper Mulberry
Broussonetia papyrifera (L.) Vent.

A tree with a broad, rounded crown when growing in the open, and with smooth, light-colored bark and milky sap. A native of Japan and China, planted about homes and now naturalized in and around cities and towns. Twigs yellowish brown to grayish, covered with hairs visible under a 10x hand lens. Leaves usually alternate, occasionally opposite; on long petioles, the blades up to 8 inches in length, roughly ovate, sometimes lobed on one or both sides, tapering to an extended tip, the margins toothed, and the base rounded to slightly asymmetric; the upper surface rough to the touch, and the lower velvety. Flowers minute, greenish, male and female on different trees, the male in narrow, hanging clusters up to 3 inches long; the female in compact spheres about ¾ inch wide, appearing in spring. Fruit fleshy, composed of many small, red fruits clustered together. Closely resembling red mulberry except for the fruit.

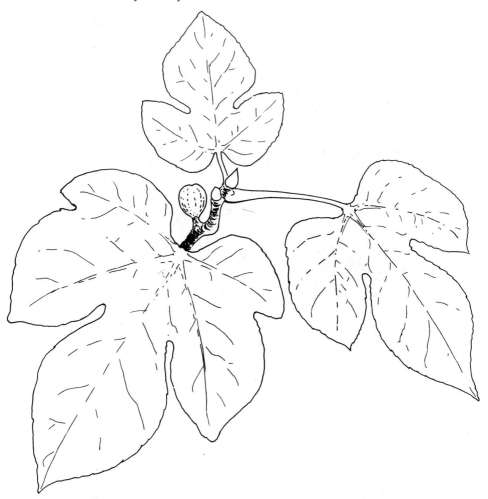

Common Fig *
Ficus carica L.

A large, broad shrub, often with spreading branches close to the
ground, with thick, gray twigs, and milky sap. Introduced from
western Asia, cultivated for its fruit, and occasionally found in
valley bottoms. Leaf blades up to 8 inches in length and of equal
width, shallowly to deeply cut into 3 or more lobes, usually with
the terminal one narrowed at the base. Leaf margins mostly with
broad, low teeth. Surfaces somewhat rough to the touch on
old leaves and soft on new ones. Flowers invisible, inside a hollow,
fleshy, broadly pear-shaped structure which ripens into the edible
fig.

Mistletoe, Injerto *
Phoradendron tomentosum (DC.) Gray subsp. *tomentosum*

An evergreen shrub parasitic on trees. Widely used in the United
States as a Christmas decoration. Stems smooth, green, brittle.
Leaf blades thick, leathery, roughly elliptic, up to 2 inches long and
1⅛ inches wide, with smooth margins, a rounded tip, and tapered
base. Plants male or female, easily distinguished in winter by
shiny, white berries ³/₁₆ inch wide on the female, and spikes of
greenish yellow flowers on the male. Especially common on Sugar
Hackberry, Cedar Elm, and Honey Mesquite, also on other trees.

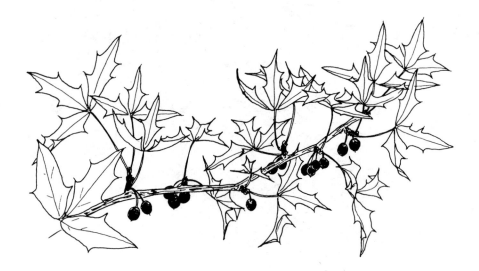

Agarito*
Berberis trifoliolata Moric.

A common shrub of old pastures and brushy areas. Leaves evergreen, firm textured, divided into three leaflets, with spine-tipped lobes. Wood bright yellow. Flowers numerous, yellow, up to ½ inch wide, appearing in February and March, their fragrance often filling the air where they are plentiful. Fruit a red berry, edible.

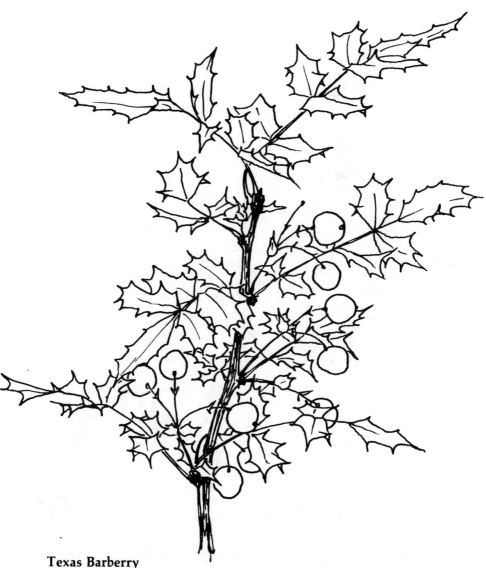

Texas Barberry

Berberis Swaseyi Buckl.

Shrub up to 3 or 4 feet tall. Local and rather uncommon on
shallow soil over limestone in and around Hays and Comal
Counties. Leaves up to 3 inches long, with 2 to 4 pairs of leaflets
and a terminal one on a central axis; leaflets firm textured,
evergreen, margins with spiny teeth, veins prominent on the lower
surface. Flowers yellow, about ⅜ inch wide, appearing from late
February to early April. Fruit a red berry, edible.

Carolina Snailseed, Red-berried Moonseed *
Cocculus carolinus DC.

A twining vine with slender, dull green stems covered with fine
hairs visible under a 10x hand lens. Frequent to common in
fencerows and brushy areas. Stems woody at the base, climbing on
fences and shrubbery and into the lower branches of trees. Leaves
triangular to heart shaped, often with a broad lobe on each side.
Flowers small, greenish, male and female on different plants, both
in loose lateral and terminal clusters, the male branched, the
female unbranched, appearing from June to August. Fruit fleshy,
bright red, ¼ inch or more in diameter. Seed coiled, suggesting a
snail.

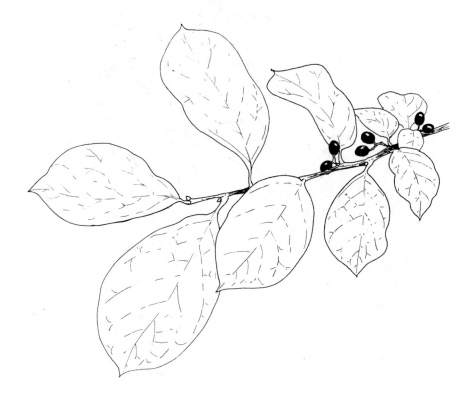

Spicebush
Lindera Benzoin (L.) Bl.

A shrub or small tree of wooded canyons. Leaves alternate on the branchlets, up to 6 inches long and 2½ inches wide, upper surface dark green, lower surface lighter in color, obovate, tapering more gradually to the base than to the tip, tip somewhat extended, margins without teeth or lobes. Flowers minute, without petals, yellow, appearing in February before the leaves. Fruit fleshy, red, less than ½ inch long, single or in clusters.

Mock-orange
Philadelphus Ernestii Hu

Small shrub with loose bark on the main stems. Among boulders
along streams. Rare and needing protection. Leaves narrowly
ovate, up to 1⅜ inches long by ½ inch wide, with smooth margins
and 3 prominent veins on the upper surface, the lower surface
covered with appressed hairs visible under a 10x hand lens.
Flowers white, showy, numerous along the branchlets, opening in
April and May. Fruit a capsule shaped like a top, 3/16 inch long by
3/16 inch wide with 4 remnants of the flower projecting from the
midpoint between the tip and the base.

Sycamore, Buttonwood, Plane-tree *
Platanus occidentalis L.

A tree of stream beds and banks, growing to a height of over 90 feet in canyons. Sometimes planted about homes. Bark scaling off in thin plates, resulting in a mottled brown and buff trunk and limbs. Leaves broadly ovate or broader, blade often wider than long, long pointed. Fruiting structures spherical, 1 inch in diameter, on long stems, persistent on the branches after the leaves have fallen.

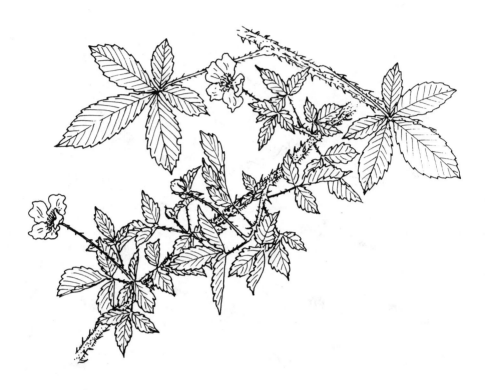

Southern Dewberry *
Rubus trivialis Michx.

Trailing shrub with recurved spines on the canes, prickly. Forms
patches along roadsides, at the edges of thickets, and in fields.
Leaves divided into 3 or 5 leaflets with toothed margins. Flowers
white, up to 1 inch wide, appearing in March and April. Fruit a
blackberry.

Hawthorn   Not Illustrated
Crataegus spp.

Species closely related and difficult to identify, hence treated
collectively here. Infrequent in the Hill Country. Shrubs or small
trees, usually with long, sharp thorns. Leaves variable in size and
shape. Flowers in clusters, white, with 5 petals, attractive, opening
in spring. Fruit fleshy, round, with a remnant of the flower
attached to the end away from the stem, resembling a rose hip.

Macartney Rose
Rosa bracteata Wendl.

Planted in fencerows and escaping into fields; forms impenetrable, prickly thickets. Leaves with 3 to 4 pairs of leaflets and a terminal one on a central axis. Leaflets firm textured, shiny above, narrowing more gradually to the base than to the rounded tip, margins with small, rounded teeth. Flowers white, showy, up to 3 inches wide, with 5 broad petals. Fruit spherical, orange red, with sepals (parts of the flower) attached at the tip. Native of China.

Laurel Cherry, Carolina Cherry-laurel
Prunus caroliniana (Mill.) Ait.

A tree with smooth, gray bark. Native of East Texas. Planted as an
ornamental in Austin and found occasionally in wooded
bottomlands along streams as a shrub or small tree. Leaves firm
textured, evergreen, narrowly elliptic, tapered to a pointed tip and
equally tapered to the base, margins smooth on reproductive trees,
with narrow, pointed teeth on saplings and rootsprouts, upper
surface dark green and shiny, lower surface lighter and duller,
having a taste suggestive of almond flavoring, poisonous when
eaten. Flowers white to cream, about 3/16 inch wide, in showy
elongate clusters among the leaves, opening from February to
April. Fruit fleshy, but with a thin pulp, black, ½ inch long by ⅜
inch wide, egg shaped with a small tip, persistent through winter.

Black Cherry, Escarpment Black Cherry*
Prunus serotina Ehrh.

A tree of canyons and wooded limestone slopes. Bark of limbs and branches reddish brown with gray or white stripes perpendicular to their axes. Leaves shiny on the upper surface; blade oblong with a long-pointed tip and tapering base, margins finely serrate. White flowers in elongate clusters in April and May followed by small cherries that are often bitter.

Texan Almond
Prunus minutiflora Engelm.

Shrub up to 3 feet tall, often thicket forming, bark gray, new
growth on branchlets covered with woolly hairs. Rare in the
Austin area. Leaves up to ¾ inch long, firm textured, oblong to
elliptic, with margins usually without teeth, a rounded, sometimes
pointed, tip, and gradually tapered base. Flowers small, white, on
short spurs, appearing in March with the leaves. Fruit spherical,
about ½ inch long, fleshy but mostly pit, black when ripe. On soils
underlain with limestone and on limestone slopes and ledges.

Mexican Plum*
Prunus mexicana Wats.

A small tree on limestone and on stream banks. Leaves up to 5 inches long and 2 inches wide, ovate to narrower with serrate margins; minute glands on the petiole near the base of the blade. Flowers white, up to ¾ inch wide, borne in flat clusters of 2-4, appearing in early spring before the leaves have reached full size. Fruit a purplish red plum.

Creek Plum, Hog Plum
Prunus rivularis Scheele

A thicket-forming shrub on stony upland sites, in wooded canyons,
and in valley bottoms. Leaves up to 2½ inches long, ovate to
narrower, with small gland-tipped teeth on the margins. Flowers in
clusters of 2 to 4 along the branches, white, up to ½ inch wide;
very noticeable in early spring in a drab countryside before many
woody plants have put out new leaves. Fruit fleshy, as much as ¾
inch in diameter, yellow to bright red or crimson.

Huisache, Sweet Acacia
Acacia Smallii Isely

A round-topped small tree of open fields, often with several trunks
and shrublike. At the northern limit of its range in the Austin
area. Branchlets spiny and bearing finely divided leaves, each of
the many leaflets less than ¼ inch long. Flowers bright yellow,
fragrant, arranged in spherical clusters ½ inch in diameter,
appearing in April. Fruit a reddish brown to black woody pod 1½
to 3 inches long, rounded, not flat, and tapered at both ends.

Catclaw Acacia
Acacia Roemeriana Scheele

A prickly shrub, occasionally a small tree. Infrequent in the Austin area, more frequent west and south. Branchlets with scattered straight or recurved prickles. Leaves divided into numerous leaflets ½ inch long or less. Flowers small, whitish, in spherical clusters ⅜ inch wide, arising mostly from the axils of the leaves. Fruit a brown to reddish, flat, curved pod up to 4 inches long and 1⅛ inches broad.

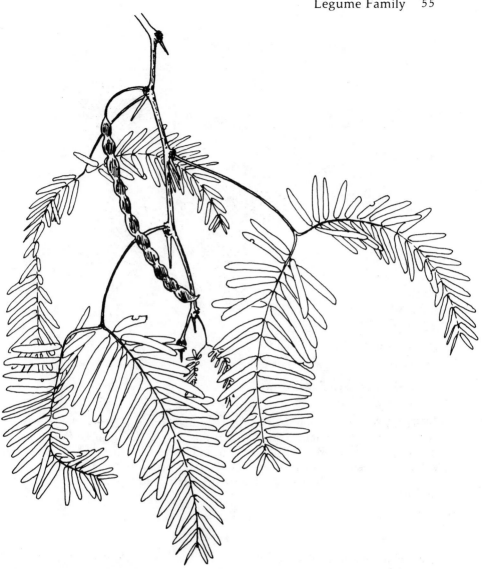

Honey Mesquite*
Prosopis glandulosa Torr. var. glandulosa

A small tree with low, drooping branches, also a shrub.
Widespread in fields out of cultivation and abandoned pastures.
Branches with spines up to 2 inches long. Leaves divided into
narrow leaflets up to 2 inches long and 3/16 inch wide. Flowers
small, creamy white, borne on spikes appearing in April and as late
as August during wet summers. Fruit a long, yellowish brown pod,
somewhat flattened and with slight constrictions between the
seeds.

Pink Mimosa, Fragrant Mimosa*
Mimosa borealis Gray

A prickly shrub of open, brushy places. Leaves twice pinnately compound, with usually 1 pair, sometimes 1 to 3 pairs, of leaflet-bearing axes, leaflets up to ⅜ inch long. Flowers in fluffy, pink, spherical clusters ½ inch or more wide, appearing in April. Fruit a flat pod up to 2 inches long, often with recurved prickles on its margins.

Cat's-claw Mimosa, Wait-a-bit   Not illustrated
Mimosa biuncifera Benth.

A shrub closely resembling Pink Mimosa. Occasional in open, unshaded areas. Branchlets armed with recurved prickles that catch on clothing. Leaves twice pinnately compound with up to 8 pairs of leaflet-bearing axes, and as many as 12 or more pairs of leaflets per axis. Leaflets about ⅛ inch long. Flowers small, pinkish, in fluffy, spherical clusters, appearing in April. Fruit a dry, flat pod, light brown to reddish brown, up to 1½ inches long, with recurved prickles on the edges.

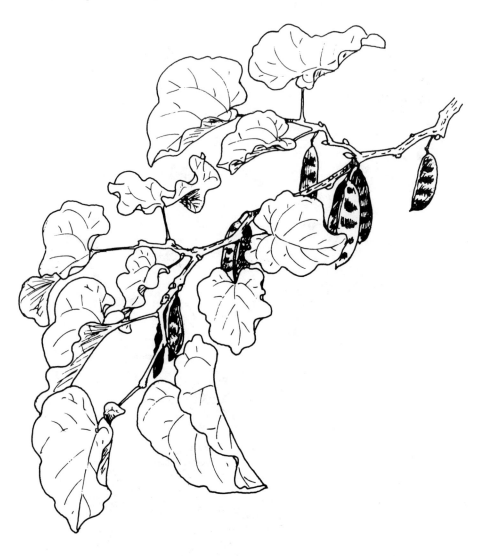

Redbud*

Cercis canadensis L. var. *texensis* (Wats.) Rose

Brightens the early spring landscape with its mass of flowers.
Leaves heart shaped to kidney shaped, rounded at the tip. Flowers
rose purple, in small clusters along the branches, appearing before
the leaves, in March or early April. Fruit a reddish brown flat pod
up to 4 inches long and pointed at the tip.

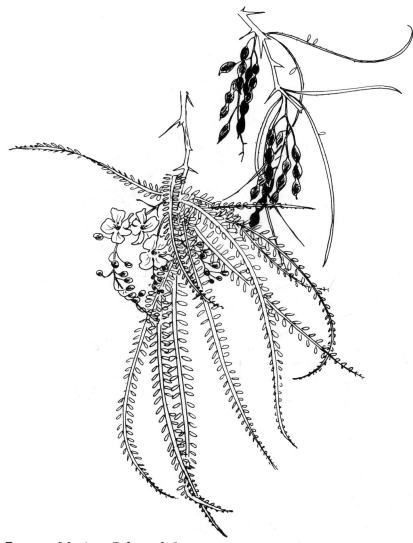

Retama, Mexican Paloverde *
Parkinsonia aculeata L.

A small tree with green bark, turning to brown on older wood, and having spiny branches. Near the northern limit of its range in the Austin area but spreading in fields out of cultivation and on disturbed ground. Leaves somewhat plumelike, finely divided into numerous small leaflets on long, flat, green axes. Flowers bright yellow, borne in loose clusters among the leaves. Fruit a pod 2 to 4 inches long, with flattened constrictions between the seeds, and tapered at both ends.

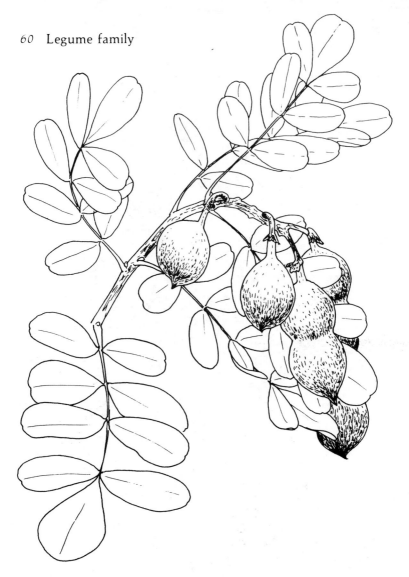

Texas Mountain Laurel, Mescal Bean *
Sophora secundiflora (Ort.) DC.

An evergreen shrub with firm-textured leaflets shiny on the upper surface. Found on rocky slopes and in shallow soil over limestone, often planted as an ornamental. Leaflets up to 2 inches or more long, tapering more gradually to the base than to the tip, and arranged along an axis terminated by a single leaflet. Flowers bluish purple, showy, fragrant, in thick clusters as much as 5 inches long, appearing in March and April. Fruit a hard pod up to 5 inches long, rounded and constricted between the red seeds, sometimes with one seed. Flowers and seeds poisonous.

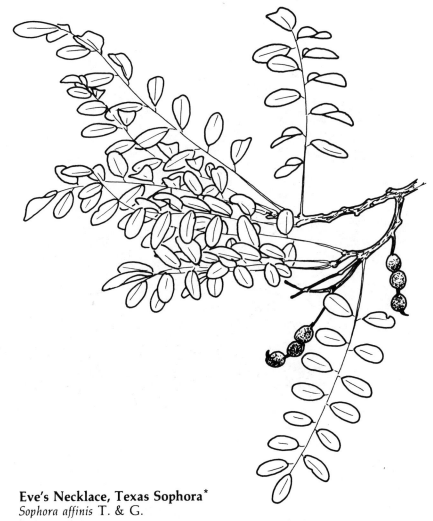

Eve's Necklace, Texas Sophora*
Sophora affinis T. & G.

A tall shrub or small tree with thin, scaly, reddish brown bark on older wood and with smooth twigs. On limestone slopes, in valley bottoms, and on soils underlain with limestone in upland situations. Seeds reputed to be poisonous. Leaves divided into 6 to 8 pairs of leaflets and a terminal one on an axis up to 9 inches long, leaflets elliptic to oval, averaging an inch long, with a rounded, indented, or pointed tip, smooth margins, and a rounded or tapered base. Flowers fragrant, white tinged with rose, ½ inch long, arranged along axes up to 6 inches long, appearing in March and April. Fruit a long, rounded pod, constricted between the seeds, often with only 1 or a few seeds, the swollen part of the pod black, and the constrictions covered with gray hairs.

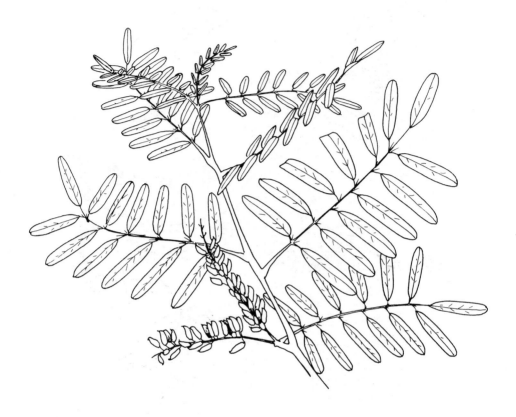

Indigo Bush, False Indigo*
Amorpha fruticosa L. var. angustifolia Pursh

Shrub commonly found in light shade on high stream banks. Leaflets 1½ inches or more long, arranged along an axis up to 8 inches long. Leaflets velvety on the lower surface, margins frequently almost parallel, often abruptly rounded at both ends and with a notch at the tip. Flowers small, purple to dark blue, with yellow stamens extending beyond the single petal, crowded in narrow, spikelike clusters at or near the ends of the branchlets, appearing from April to June. Fruit small, up to ⅜ inch long and with blisterlike glands visible under a 10x hand lens.

Texas Kidneywood [*]
Eysenhardtia texana Scheele

A medium to tall shrub common on limestone uplands in the open and in light shade. Leaves up to 3½ inches long, consisting of a central axis and as many as 40 small leaflets, each about ¼ inch long, pungent when crushed. Flowers white, small, with a delicate fragrance, arranged in spikes up to 4½ inches long at the ends of branchlets, appearing intermittently from May to October. Fruit a pod about ⅜ inch long, often with a threadlike tip.

Black Dalea*
Dalea frutescens Gray

Shrub up to 3 feet tall; stems gray to light brown, leaf-bearing
twigs thin, reddish brown. Occasional on shallow soil over
limestone in unshaded upland situations. Leaves up to 1 inch long,
divided into as many as 8 pairs of small leaflets and a terminal one
on a central axis, leaflets 5/16 inch or less long, gland dotted on
the lower surface. Glands visible under a 10x hand lens. Flowers in
dense heads or spikes at the ends of branches, small, purple,
opening from July to October. Fruit an inconspicuous capsule.

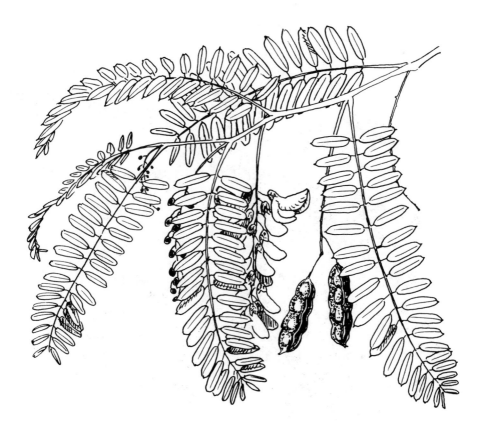

Rattlebush, Rattlebox
Sesbania Drummondii (Rydb.) Cory

An erect shrub up to 8 feet tall of unshaded, low, wet places,
frequently freezing back to the base during the winter and giving
the impression of being a fast-growing annual. Bark smooth, green
to light brown. Leaves up to 8 inches long, usually around 6
inches, with paired leaflets openly spaced along a central axis and
with no terminal leaflet. Leaflets up to 1¼ inches long, narrow,
with a short tip. Flowers about ½ inch long, yellow, often streaked
with red, arranged in loose elongate clusters. Fruit a 4-angled or
winged pod up to 2½ inches long, slightly constricted between the
seeds and tapering to a beak at the apical end, rattling when dry.

Black Locust
Robinia pseudo-acacia L.

An introduced tree, native to the southeastern United States. Branches and twigs with spines. Leaves divided into ovate to oblong leaflets up to 2 inches long and 1 inch wide, rounded at the ends and with smooth margins. Flowers appearing in April and May, white, the upper petal often with a yellow area on its inner surface, fragrant and in showy pendulous clusters 4 inches or more long. Fruit a flat, straight to slightly curved pod up to 5 inches long.

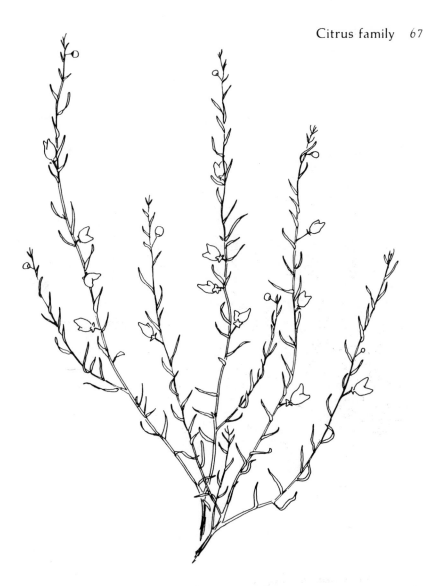

Texas Desert-rue, Ruda del Monte, Dutchman's Breeches
Thamnosma texana (Gray) Torr.

A low, aromatic plant with a woody base and thin, herbaceous stems up to 1 foot tall. Grows in open and brushy areas in shallow soil over limestone, more common west of the Hill Country. Leaves small, narrow, up to ¾ inch long by ¹/₁₆ inch wide, dotted with glands visible under a 10x hand lens. Flowers small, ⅛ to ³/₁₆ inch wide, arranged along the upper half of the stems, appearing from March to June. Fruit a 2-lobed capsule having an inflated appearance and giving rise to the name "Dutchman's Breeches."

Wafer-ash, Common Hop-tree *
Ptelea trifoliata L.

A shrub of stream terraces and meadows, and at the edges of
thickets on upland sites. Resembles a small ash, but not related.
Leaves with 3 leaflets on a petiole up to 2 inches long, the terminal
leaflet up to 2½ inches long, obovate, tapering more gradually to
the base than to the tip, midrib of lateral leaflets off-center, leaves
yielding a fetid odor when crushed. Flowers small, greenish white,
in clusters among the leaves, appearing in April. Fruit distinctive,
waferlike with broad wings, approximately ⅞ inch long by ¾ inch
wide.

Prickly Ash, Toothache Tree, Tickle-tongue*
Zanthoxylum hirsutum Buckl.

A prickly shrub, sometimes a small tree. Crushed leaves have an odor suggestive of orange peel. Found in open areas and at the edges of woodlands. Branches with prickles resembling those of a rosebush. Leaves with paired glossy leaflets along an axis bearing small prickles. Leaflets up to 1½ inches long, the margins with glands in the notches between the rounded teeth. Flowers small, greenish, in clusters at the ends of branchlets. Fruit spherical, ¼ inch in diameter, reddish brown when ripe.

Tree-of-Heaven, Copal Tree
Ailanthus altissima (Mill.) Swingle

A fast-growing tree of waste places and disturbed ground but also planted as an ornamental. Naturalized from Asia. Bark rather smooth, not furrowed. Branches with a large reddish pith and with heart-shaped scars left by fallen leaves. Leaves large, with a central axis up to 2 feet long and the blade divided into as many as 20 pairs of leaflets, usually fewer, with or without a single terminal one. Leaves give an unpleasant odor when crushed. Flowers very small, in large, loose clusters. Fruits dry and winged with the seed in the center.

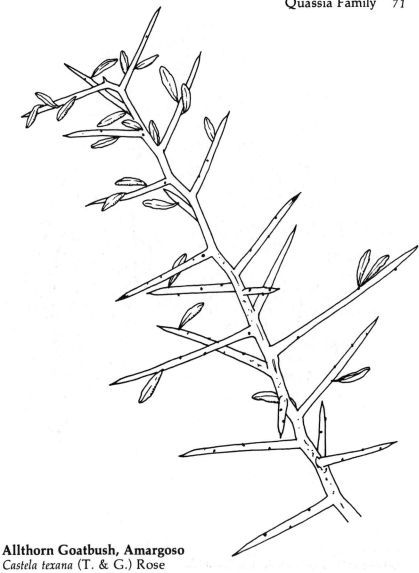

Allthorn Goatbush, Amargoso
Castela texana (T. & G.) Rose

A thicket-forming, spiny shrub with bitter bark. Found on exposed
hillsides and in prairies. Branchlets light colored, terminating in
stout spines bearing lateral ones. Leaves firm textured, up to 1
inch long by ¼ inch wide, with a pointed or rounded tip and
smooth, turned-down margins, the upper surface shiny, the lower
covered with fine hairs visible under a 10x hand lens. Flowers
small, red to orange, occuring singly or in small groups. Fruit
fleshy, red, roughly spherical, slightly flattened, up to ⅜ inch long.
Bark reputed to have medicinal properties.

Chinaberry, Pride-of-India, Canelón*
Melia Azedarach L.

A fast-growing tree of disturbed sites, thickets, and lowland soils; planted about homes and escaped from cultivation. Naturalized from India. Bark gray to reddish brown. Leaves usually a foot or more long, twice pinnately compound, the leaflets on the lateral axes ovate to narrower, with toothed margins and pointed tips. Flowers with purple centers and lavender to white petals; in large, open sprays, fragrant, appearing in April. Fruit spherical, ½ inch or more in diameter, yellow, fleshy, and with a hard pit.

Narrow-leafed Thyrallis
Thyrallis angustifolia (Benth.) O. Ktze.

A low shrub occasional in stony, well-drained places, in the open
and in light shade, in the Hill Country. Seldom over 18 inches tall
with erect stems 1/16 inch in diameter arising from a woody base;
the stem, except near the base, green and herbaceous. Leaves
opposite, somewhat distantly spaced on the stem, narrow, up to
1¾ inches long by 3/16 inch wide, with a smooth, often rolled-down
margin. Flowers about ½ inch wide, yellow to orange and red,
scattered along the upper 1/3 of the stem, above the leaves,
appearing from early March to July. Fruit a small, 3-lobed capsule
about 3/16 inch long.

Maidenbush
Andrachne phyllanthoides (Nutt.) Coult.

A low shrub, usually less than 3 feet tall, with arched, leafy stems.
On limestone cliffs along streams; apparently rare shrub. Leaves
numerous, small, up to ½ inch long by ⅜ inch wide, tapering more
gradually to the base than to the rounded tip. Flowers greenish,
not showy, on threadlike pedicels roughly ½ inch long arising from
the axils of the leaves. Male and female flowers not the same size,
male about ¼ inch wide and female almost twice this size, and on
separate plants, appearing from April through summer. Fruit a
small, rounded capsule broader than long.

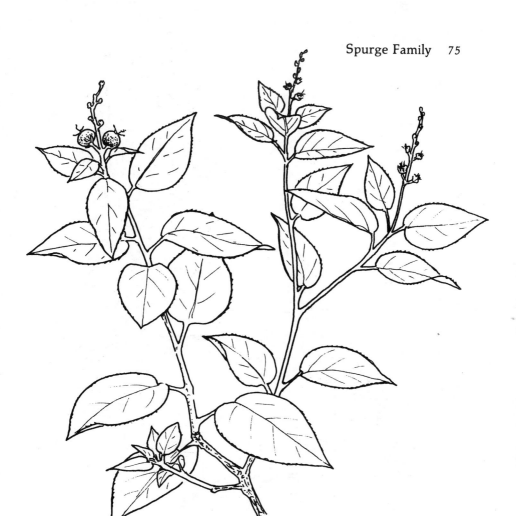

Encinilla, Bush Croton*
Croton fruticulosus Torr.

A shrub up to 3 feet or more tall. On limestone ledges and slopes
west of the Balcones Escarpment. Stems slender, smooth, brown
or gray, leaf-bearing branchlets yellowish or grayish green and
densely covered with stellate hairs visible under a 10x hand lens.
Leaf blades up to 3 inches long, ovate to narrower, with smooth to
sparsely and finely toothed margins, pointed at the tip, and
rounded or slightly lobed at the base. Both surfaces on younger
leaves velvety to the touch, the upper frequently olive green, the
lower grayish green and densely covered with stellate hairs,
aromatic when crushed. Flowers about ⅛ inch wide, on terminal
extensions of the branchlets. Fruit a slightly 3-lobed spherical
capsule ¼ inch in diameter, covered with stellate hairs.

Brush Myrtle-croton, Oreja de Ratón
Bernardia myricaefolia (Scheele) Wats.

A thickly branched shrub with gray bark and numerous spurs
bearing small leaves. Found on rocky slopes of canyons and hills,
near the northeastern limit of its range in the Austin area. Leaves
more or less elliptic, dark green on the upper surface and grayish
green on the lower, with rolled-down, toothed margins. Blades in
full sun generally less than ⅝ inch long, gray and densely hairy on
the lower surface, and appearing to be twisted or distorted; blades
in the shade up to 2 inches long, flat and light green on the lower
surface. Flowers inconspicuous. Fruit a 3-lobed spherical capsule
about ⅜ inch in diameter.

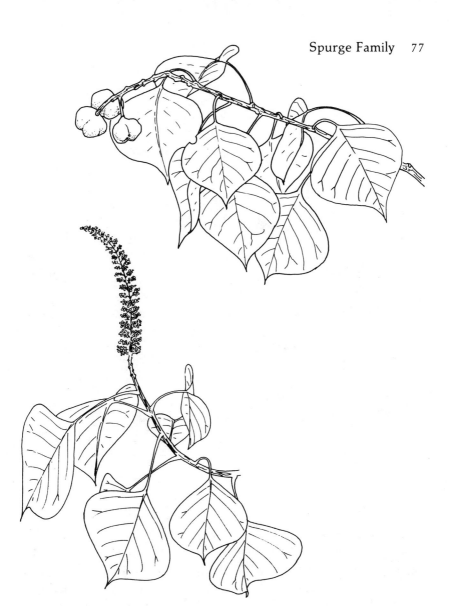

Chinese Tallow Tree*
Sapium sebiferum (L.) Roxb.

A fast-growing, popular lawn tree that has escaped cultivation. Naturalized from China. Leaves with long stems, blades roundish with smooth margins and tapering to a long tip, reddened in autumn. Flowers small, yellowish green, arranged in spikes with the male borne above the female. Fruit a capsule with 3 white seeds that persist on the tree long after the walls of the capsule have fallen away.

Poison Ivy, Poison Oak *

Toxicodendron radicans ssp. *verrucosum* (Scheele) Gillis

A shrub or climbing vine, older stems of vines fastened to trees by a dense growth of aerial rootlets and often climbing into treetops. Common to abundant in and at the edges of wooded areas, and containing an oil capable of causing intense itching and watery blisters through contact with the skin. Leaves variable in size, divided into 3 leaflets on an often reddish tinged petiole; the terminal leaflet symmetric, roughly ovate but lobed and toothed, larger than the lateral ones; lateral leaflets asymmetric, lobed and toothed on the side toward the leaf base and usually only toothed on the opposite side, turning orange and red in autumn. Flowers greenish white, inconspicuous, in clusters up to 4 inches long. Fruit cream to white, sometimes a dull yellow, spherical, about ¼ inch in diameter, conspicuous on bare twigs in winter.

Prairie Flameleaf Sumac *
Rhus lanceolata (Gray) Britt.

A tall, openly branched shrub or small tree, often in clumps. Common in openings in cedar-oak woodlands, in abandoned pastures, and in old fields out of cultivation. Bark gray to brown, smooth to scaly with age, new growth covered with velvety hairs. Leaves up to 14 inches long with the blade divided into as many as 21 narrow, pointed leaflets arranged in pairs along a winged axis with a terminal leaflet. Leaflets asymmetric with the wider part of the blade toward the leaf tip, often with the midrib curved slightly toward the base, brightly colored in autumn. Flowers small, whitish, numerous, in clusters up to 5 inches long and 3 inches wide at the tips of branches, followed by densely clustered red fruits. Fruit clusters distinctive, persisting on the branches after the leaves have fallen.

Evergreen Sumac *
Rhus virens Gray

Shrub or small tree. Frequent to common on rocky slopes and in light shade in cedar-oak woodlands. Leaves evergreen, up to 6 inches long, with a central axis and 4 to 8 paired leaflets and a terminal one; leaflets firm textured, dark green and shiny on the upper surface, dull on the lower, oval to narrower, with a rounded or broad tip; the terminal one with a wedge-shaped base, the laterals less so, up to 1½ inches long and ¾ inch wide, a few larger. Flowers minute, whitish, in clusters up to 2 inches long among the leaves near the tips of branches. Fruit ¼ inch long, oblong and somewhat flattened, red, and covered with short hairs.

Fragrant Sumac, Lemon Sumac, Polecat Bush *
Rhus aromatica Ait. var. *flabelliformis* Shinners

A frequent shrub in fencerows, brushy areas, and in light shade under trees. Bark smooth, light brown to reddish brown. Leaves of variable size, dark green in summer, brightly colored in the fall, with 3 leaflets; terminal leaflet typically 3-lobed, tapering more than half its length to a wedge-shaped base; lateral leaflets smaller than the terminal one and usually not as lobed nor as wedge shaped. Flower buds in winter appearing as small reddish spikes at the tips of branches. Branches give odor when broken. Flowers small, yellow, in tight spikelike clusters, opening in March and April. Fruit small, spherical to somewhat flattened, fleshy, reddish, covered with short hairs.

Possumhaw, Winterberry, Deciduous Holly *
Ilex decidua Walt.

Tall shrub or small tree with smooth light gray bark. Found on brushy and wooded slopes and in low, wooded places. Leaves in clusters on short spurs or arranged alternately on the branchlets; up to 3 inches long but more commonly half this length. Leaves tapering gradually to a petiole and suggesting a broad paddle blade, rounded at the tip and with rounded teeth tipped with minute glands. Gland visible under a 10x hand lens. Flowers small, white, mostly clustered at the tips of the leaf-bearing spurs, opening from March to May. Fruit fleshy, spherical, about ¼ inch in diameter, bright red, persisting after the leaves have fallen.

Yaupon*
Ilex vomitoria Ait.

Evergreen shrub or small tree with smooth, tight bark. Found in
wooded bottomlands and wooded canyons. Twigs stiff, grayish,
sometimes branching almost at right angles from the stout
branchlets. Leaf blades firm textured, dark green and lustrous on
the upper surface, lighter green and duller on the lower surface;
up to 2¼ inches long but mostly shorter; ovate to oblong. Margins
with rounded teeth tipped with a gland visible under a 10x hand
lens. Flowers small, white, appearing in April and May. Fruit
fleshy, about ¼ inch in diameter, bright red.

Boxelder, Ash-leaved Maple, Fresno de Guajuco
Acer Negundo L.

Found occasionally on stream banks. Twigs green to purplish and smooth. Leaves divided into 1 to 4 pairs of unsymmetrical, irregularly toothed, sometimes shallowly lobed, leaflets on a central axis and a symmetrical terminal leaflet. Flowers greenish, small, without petals, appearing before the leaves. Fruits in pendulous clusters, dry, winged and paired, with the seeds at the point of attachment.

Red Buckeye*
Aesculus Pavia L. var. *Pavia*

An attractive spring-flowering shrub or small tree, blooming in
some years as early as February, with showy red flowers
sometimes partly yellow. Frequent on limestone ledges above
streams. Sometimes showing hybridization with the yellow-
flowered var. *flavescens* (Sarg.) Correll, the common form westward
on the Edwards Plateau. Large leaves with leaflets radiating from
the tip of a long petiole. Bark on young branches smooth, gray to
brown, roughened on older ones. Leaf blade palmately divided into
5 leaflets; petiole up to 6 inches long and leaflets of equal length,
tapering more gradually to the base than to the elongate tip, and
with serrate margins. Flowers tubular, 1¼ inches long, on an
upright axis as much as 8 inches tall. Fruit a rounded capsule 2
inches in diameter, brown, with a slightly roughened surface,
persisting after the leaves have fallen; seeds 1 to 3, shiny.

Western Soap-berry, Jaboncillo *
Sapindus Saponaria L. var. *Drummondii* (H. & A.) L. Benson

A small to medium-sized tree at the edges of woodlands and in valleys. Bark rough, flaking off in chips, gray to light reddish. Leaves up to 18 inches long with a central axis and as many as 24 paired leaflets, usually fewer, and often no terminal leaflet. Leaflets unsymmetric with the broader part of the blade toward the leaf tip and the base rounded on the broader side and tapering on the narrower side. Leaflet tip elongate. Flowers in large, cream-colored clusters up to 10 inches long and 6 inches wide, appearing in May and early June. Fruit fleshy, globose, about ½ inch wide, flesh transluscent, yellow turning darker with age, sometimes persistent on the tree until the next flowering season.

Mexican Buckeye, Texas Buckeye , Monilla*
Ungnadia speciosa Endl.

Shrub or small tree with light gray to brown bark, smooth on young branches, becoming fissured with age. Occasional near streams and on wooded limestone slopes. Leaves up to 12 inches long, with a central axis supporting 2 to 6 paired leaflets and a terminal one; leaflets up to 5 inches long, ovate to narrower with an elongate tip, rounded base, and serrate margins. Flowers pink, tinged with purple, showy, delicate, fragrant, in clusters near the ends of the branches, appearing before or with the expanding leaves. Fruit distinctive, a light reddish brown when ripe, 3-lobed capsule containing 1 to 3 dark brown to black, shiny seeds ½ inch in diameter, the walls of the capsule often persisting through the winter, seeds poisonous.

Texas Colubrina, Hog-plum *
Colubrina texensis (T. & G.) Gray

A thicket-forming shrub. Found on shallow, rocky soil. More common southward. Twigs branching almost at right angles from the branchlets. Leaves clustered in alternate bundles. Leaves also found singly on young growth. Leaves up to 1 inch long, but often about half this length, and from less than ¼ to ½ inch wide, finely serrate and soft to the touch, tapered more gradually to the base than to the tip. Flowers small, greenish yellow, on short pedicels arising from the same points as the leaves, opening in April and May. Fruit brown to black, a little over ¼ inch in diameter, rounded, with a short beak.

Redroot, Inland Ceanothus
Ceanothus herbaceus Raf.

A low shrub, usually less than 3 feet tall, with a thick rootstock. Occasional on stony soil in fields and pastures and in light shade in cedar-oak woodlands. Leaves as much as 2½ inches long and 1 inch wide, may be as small as 1 inch by ¼ inch, with prominent, yellowish veins on the lower side. Margins finely serrate. Flowers small, white, in dense, rounded clusters ½ to ¾ inch wide, at the ends of leafy twigs, opening from March to July. Fruit a rounded, dark brown, 3-lobed capsule, about 3/16 inch in diameter, with a saucerlike support.

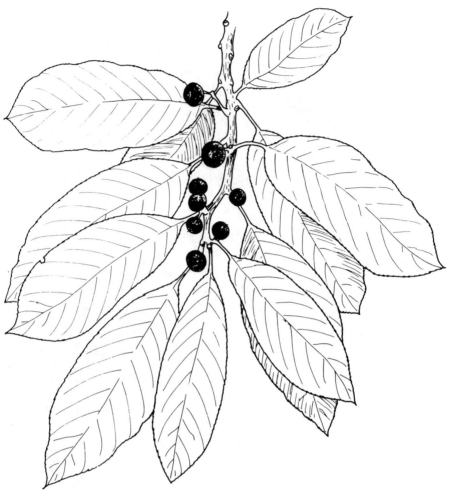

Indian-cherry, Carolina Buckthorn*
Rhamnus caroliniana Walt.

Tall shrub or small tree with leaves that stay green into late fall.
Frequent near streams and also on wooded and brushy slopes.
Leaves up to 5 inches long, with a petiole as much as ½ inch long;
blade ovate to elliptic, sometimes narrow, pointed at the tip and
tapered or rounded at the base, margins smooth or with very
small, rounded teeth, veins prominent, especially on the lower
surface; upper surface of blade smooth, bright green. Flowers not
showy, yellowish, in small clusters at the bases of the leaves,
opening in May and June. Fruit fleshy, ¼ inch or more in diameter,
red, turning black when ripe.

Alabama Supple-jack, Rattan-vine*
Berchemia scandens (Hill) K. Koch

A climbing vine with distinctive smooth, green bark on stems as
much as 1 inch in diameter. Climbing into trees in wooded ravines
and narrow canyons. Leaves from 1 to 2½ inches long, with
prominent veins branching from the midrib, ovate to elliptical with
smooth or slightly wavy margins and a rounded or pointed tip.
Flowers small, less than ⅛ inch wide, in elongate, open clusters on
short lateral branchlets, appearing in April and May. Fruit small,
fleshy, blue black, longer than wide and slightly flattened.

Jujube
Ziziphus jujuba Mill.

A small tree naturalized from the Old World, rather rare in the Hill Country. Leaves on usually zigzag-shaped branchlets, ovate, shiny, with 3 prominent veins, margins shallowly toothed, tip broad, not elongate, base rounded; two spines sometimes at the base of the leaf. Flowers, in May, in small clusters at the bases of the leaves. Fruit fleshy, oblong, up to an inch or more long, brown to black when ripe, edible, with hard, elongate seeds.

Lotebush
Ziziphus obtusifolia (T. & G.) Gray

Densely branched shrub with small leaves and stiff branchlets ending in spines. Frequent in unshaded places with very shallow soil over limestone. Bark with light gray coating interrupted by fine, lengthwise cracks resulting from radial growth. Leaves in clusters or alternate, up to 1¼ inches long, but mostly about half this length, variable in shape and with serrate or smooth margins. Flowers inconspicuous, in small clusters. Fruit fleshy, black, about ⅜ inch in diameter.

Brasil
Condalia Hookeri M.C. Johnst. var. *Hookeri*

A shrub with stiff, leafy branches ending in thorns. Found on shallow soils over limestone. Branchlets of current year's growth green, older growth gray streaked with brown. Leaves bright green, paddle shaped, the widest part nearer the tip than the base, tip with a small point, margins smooth to weakly toothed toward the tip, the remainder of the blade tapering to the petiole. Flowers small, green, inconspicuous. Fruit spherical, about 5/16 inch in diameter, fleshy, black, ripening in summer.

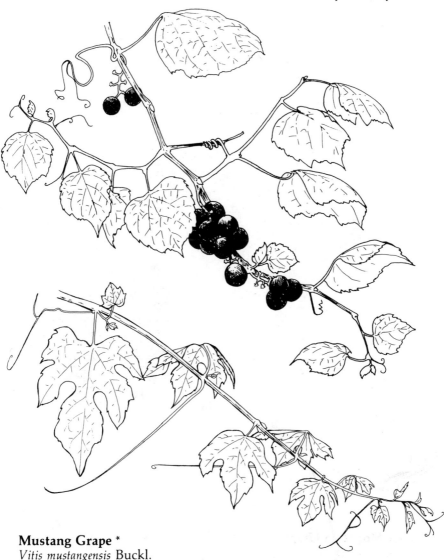

Mustang Grape *
Vitis mustangensis Buckl.

A common and easily recognized grape with a white, velvety
surface on the lower side of the leaves. A vine climbing over
shrubs and into trees and often shading their leaves. Leaves in two
forms: one form unlobed or shallowly lobed, and the other form
deeply lobed, with the latter less common and on rapidly growing
shoots. The lower surface of the unlobed leaves often concave.
Grapes up to ¾ inch in diameter, few to the bunch, ripening in
August and September to dark purple, and usually bitter, even
irritating, but popular with makers of homemade wine.

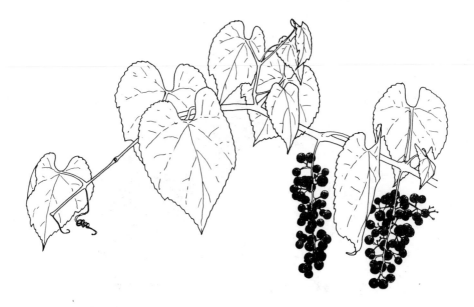

Sweet Mountain Grape
Vitis monticola Buckl.

A climbing vine growing in soils underlain with limestone west of the Balcones Escarpment. Similar to the Spanish Grape but with smaller leaves and shorter flower clusters. Leaf blades as broad as long, up to 4 inches in either direction, but usually less, roughly triangular to broadly heart shaped. Flowers in typically forked clusters up to 3 inches long, blossoming in May and June. Grapes up to ½ inch in diameter, in compact bunches, ripening in August and September, sweet tasting. Commonly black when completely ripe. But also reddish forms.

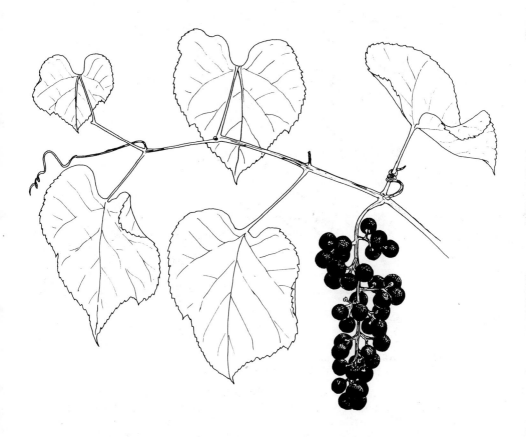

Spanish Grape, Winter Grape *
Vitis Berlandieri Planch.

A common grape with a smooth surface on the lower side of the leaves. Lower surface of mature leaves with white, cobweblike hairs restricted to the veins and visible under a 10x hand lens. Leaf blades up to 4½ inches long and 5 inches wide, broadly ovate to roughly heart shaped, usually with 2 broad lobes, a pointed tip, and broad teeth. Flowers minute, greenish, in branched, sometimes forked, clusters up to 8 inches long, appearing in April and May. Purplish to reddish grapes up to 5/16 inch wide, ripening from August to October, palatable.

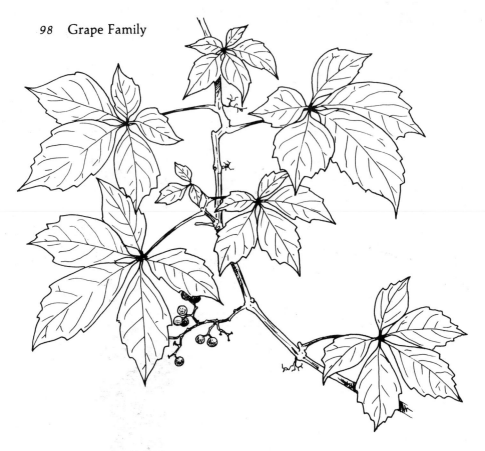

Virginia Creeper*
Parthenocissus quinquefolia (L.) Planch.

A common tendril-bearing vine climbing high into trees or trailing over the ground; tendrils with disks that fasten onto bark or rock. Leaves with 5 leaflets, occasionally 3 or 7, radiating from the tip of the petiole, coarsely toothed, with a pointed tip, and tapered to the base, up to 6 inches long, and often turning bright red in autumn. Flowers small, greenish, in clusters, appearing in spring. Fruit bluish, about ¼ inch in diameter.

Seven-leaf Creeper*   Not Illustrated
Parthenocissus heptaphylla (Buckl.) Small

Vine found occasionally west of the Balcones Escarpment on soil underlain with limestone. Resembling Virginia Creeper but much less common; tendrils without disks, leaflets narrower, thicker, and shorter than in Virginia Creeper, up to 2½ inches long, and turning color later in the fall. Leaflets usually 7 in number, but may be 5 or 6.

Heart-leaf Ampelopsis
Ampelopsis cordata Michx.

A tendril-bearing vine climbing into trees in stream bottoms and wooded canyons, bearing a resemblance to wild grapes but the fruit not edible. Leaf blades up to 5 inches long and 4 inches wide, broadly ovate, with an extended tip, coarsely toothed margins, and a truncate to heart-shaped base. Flowers inconspicuous, borne in clusters opposite the leaves, the flower stems frequently twining as tendrils. Fruit a berry up to 5/16 inch in diameter, slightly wider than long, ripening from late summer into fall.

Pepper-vine *
Ampelopsis arborea (L.) Koehne

A vigorous vine of wet areas and stream courses, growing over shrubs and into trees. Leaves up to 6 inches or more long and equally wide, with a central axis and 1 to 3 pairs of lateral axes supporting leaflets. Leaflets roughly ovate, coarsely toothed, dark green on the upper surface, lighter on the lower. Flowers inconspicuous, in small clusters opposite the leaves. Fruit fleshy, up to ⅝ inch in diameter, black and shiny when ripe, inedible.

Ivy Treebine, Cow-itch, Hierba del Buey *
Cissus incisa (Nutt.) Des Moul.

A mostly herbaceous vine, large ones woody at the base, with thick, fleshy leaves yielding a fetid odor when crushed. Climbing on trees, buildings, fences, and shrubs, frequent to common. Leaves up to 3½ inches long including the petiole, commonly 3-lobed or divided into 3 leaflets with the terminal one symmetric and the laterals asymmetric. Tendrils coiling at the tip. Flowers small, in greenish flat-topped to rounded clusters up to 2 inches wide. Fruit broadly ovate, ¼ to 1/3 inch long, black when ripe.

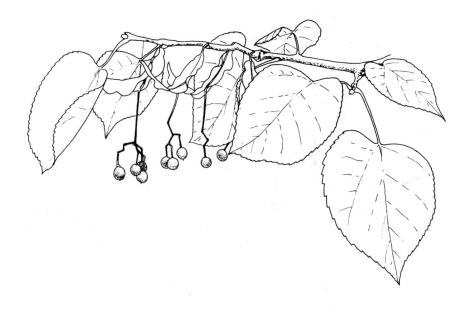

Carolina Basswood, Carolina Linden
Tilia caroliniana Mill.

Found on stream banks, infrequent. Leaves up to 5 inches long, blade broadly ovate with toothed margins, tip emerging rather abruptly from the blade, base of blade at an oblique angle to the petiole, flat to somewhat lobed. Flowers and fruits in a cluster emerging from the midpoint of a distinctive bract up to 4 inches long by an inch wide. Flowers white to yellowish; fruit spherical, about ¼ inch in diameter when mature. Very similar to *T. Floridana* Small which is also in the Hill Country.

Drummond Wax-mallow, Texas Mallow, Turk's Cap
Malvaviscus arboreus Cav. var. *Drummondii* (T. & G.) Schery

A coarse shrub, upper stems greenish and velvety to the touch, woody near the base. Occasional in light shade near streams. Leaves, including petioles, up to 5 inches or more long; blades as broad as long, broadly heart shaped to weakly 3-lobed, with broad teeth, upper surface dark green, lower surface lighter and velvety, palmately veined. Flowers showy, petals bright red, overlapping, 1 inch or more long, pistil and stamens forming a column protruding ¾ inch beyond the petals, appearing mostly in late summer and fall. Fruit red, a 5-lobed capsule cupped in green remnants of the flower.

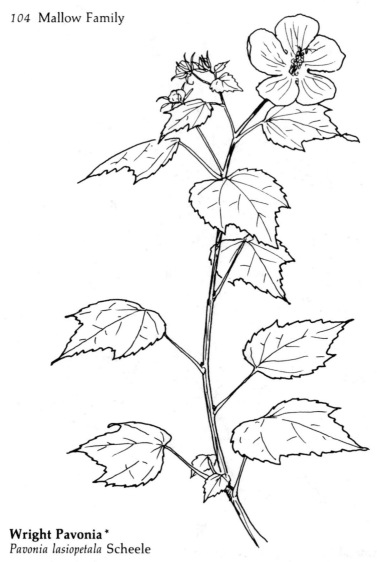

Wright Pavonia *
Pavonia lasiopetala Scheele

A small shrub, usually woody at the base only, with stems up to 4 feet tall. Found in shallow soil on limestone, in rocky places in woodlands, and at the edges of thickets. Leaves with petioles sometimes as long as the blade; blade up to 2½ inches long, but mostly shorter, ovate to 3-lobed, with a pointed or blunt tip, flat or slightly lobed base, coarsely toothed or wavy margins, dark green on the upper surface and ligher on the lower. Flowers showy, rose colored, roughly 1½ inches wide with a yellow column formed by the pistil and stamens, appearing from spring to fall. Fruit a 5-lobed capsule with remnants of the flower at its base, separating into 5 units at maturity.

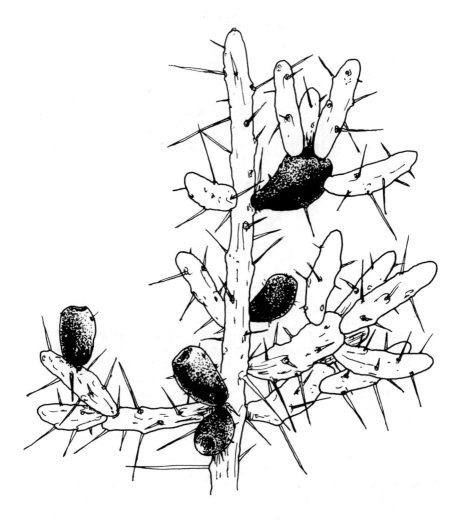

Pencil Cactus, Tasajillo, Desert Christmas Cactus *
Opuntia leptocaulis DC.

A small, leafless shrub with erect stems, sometimes leaning on other shrubs, and jointed, green, spiny branches. Occasional in fencerows and at the edges of thickets. Leaves not forming a blade and falling early in the growing season. Flowers greenish yellow, about ½ inch in diameter. Fruit fleshy, bright red, longer than wide, and ½ inch or more long.

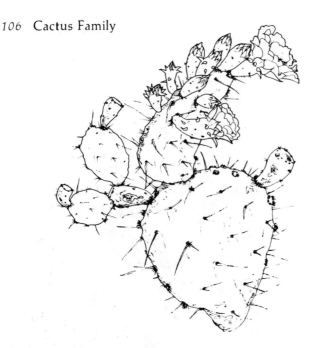

Texas Prickly Pear, Nopal Prickly Pear *
Opuntia Lindheimeri Engelm.

A common cactus in overgrazed pastures and at the edges of thickets. Stems jointed, each segment flat, green, spiny, slightly longer than broad, and covered with a thin, waxy coating. Spines in groups and of 2 types: one, ½ to 1½ inches long; and the other, about 3/16 inch long. Leaves very small, falling early in the growing season. Flowers showy, yellow to orange, appearing from late April to early June. Fruit fleshy, red or purple, rounded, flattened at the end and tapering toward the base, ripening in late summer, edible.

Brownspine Prickly Pear *   Not Illustrated
Opuntia phaeacantha Engelm

A Prickly Pear lacking an upright stem. Common to abundant in abandoned pastures and old fields on stony soil. Forms low patches of flat joints, stem segments, or horizontal lines of 3 or more joints standing on edge, some tinged reddish purple in winter. Spines of 2 kinds: one kind ½ to 2 inches long and single, or 2 or 4 together, gray to brown or yellowish, sometimes pointing downward; and the other kind minute ones in dense oval clusters from which the long spines arise. Flowers showy, yellow, often with a red center, up to 3 inches wide, opening in April and May. Fruit fleshy, up to 2¼ inches long, purplish, flattened to concave at the apex, tapering to the base.

Lindhiemer Silk-tassel *
Garrya Lindheimeri Torr.

Evergreen shrub or small tree. Frequent on limestone ledges and
rocky slopes of canyons and ravines. Twigs with leaf scars that go
completely around them and gray streaks (lenticels) running
lengthwise in the reddish brown bark. Leaves opposite, petioled;
blade leathery, variable in shape, roughly elliptic, up to 2½ inches
long, with a very small, abrupt tip and smooth margins, smooth on
the upper surface, velvety on the lower. Flowers about 3/16 inch
wide, in simple pendulous clusters from the bases of the leaf
petioles, opening in March and April. Fruit fleshy, round, with a
short tip, blue with a white coating easily rubbed off, about ⅜ inch
in diameter.

Rough-leaf Dogwood [*]
Cornus Drummondii C.A. Mey.

Shrub or small tree with smooth, reddish brown or gray
branchlets. Common near streams or ponds, usually under trees,
also on canyon slopes. Leaves opposite on green twigs, petioled;
blades up to 4 inches long, roughly ovate with an abruptly drawn-
out tip and a rounded to tapering base, smooth margins, and
prominent veins bending toward the tip; upper surface sometimes
slightly rough to the touch, lower slightly velvety. Flowers about
¼ inch wide, cream colored, with 4 petals, numerous in broad
clusters at the ends of branches, appearing from April to early
June. Fruit fleshy, rounded, white, about ¼ inch wide.

Texas Madrone *
Arbutus xalapensis H.B.K.

A small evergreen tree with smooth, pinkish bark that peels off in thin sheets, rare on limestone hills in the Austin area, more frequent westward. Petioles up to 1¼ inches long, blades to 3½ inches long, ovate to elliptic, of a leathery texture, margins usually smooth. Flowers white, small, urn shaped, in wooly clusters, appearing in early spring. Fruit spherical, up to 1/3 inch in diameter, in elongate clusters, edible.

Coma *
Bumelia lanuginosa (Michx.) Pers. var. *rigida* Gray

Shrub or small to medium-sized tree at the edges of woodlands, along streams, and in fencerows and abandoned pastures. Branchlets often with a thorn at the tip. Leaves alternate or in clusters, up to 2½ inches long and an inch wide, mostly smaller, tapering more gradually toward the base than toward the tip. Leaf margins smooth, rolled down, tip rounded, lower surface hairy. Flowers several on short stems from the bases of the leaf clusters, opening in June. Fruit fleshy, dark purple to black, elliptical, ⅜ inch long, with a remnant of the flower at the tip.

Texas Persimmon, Mexican Persimmon*
Diospyros texana Scheele

Shrub or small tree with very hard wood. Common in brushy areas on level uplands, stony hillsides, and lower slopes. Bark light gray, smooth, thin, on some trunks peeling in rectangular flakes and exposing a pinkish layer beneath. Leaves up to 2 inches long, but most about half this length, firm textured, rounded or slightly notched at the tip and tapering to the base; margins smooth, rolled down. Flowers urn shaped, whitish, about ⅜ inch wide, arranged singly or in small clusters among the new leaves; male and female on separate plants, appearing in March and April. Fruit fleshy, round, up to 1 inch in diameter, black and sweet when ripe, ripening from late July into September.

Sycamore-leaf Snow-bell
Styrax platanifolia Engelm.

A shrub or small tree with dark gray bark. Near streams in
canyons in the Hill Country. Should not be disturbed. Rare and
endangered in the vicinity of Austin. Leaves broadly ovate to
almost circular, up to 4 inches long, with smooth margins and a
broad tip, or with a terminal lobe and a lateral one on each side of
it, and a lobed-to-rounded base. Flowers in small clusters,
drooping, with 5 white petals up to ⅝ inch long, suggesting little
bells, opening in April and May. Fruit a rounded capsule about ⅜
inch in diameter, with a short tip, the base covered with a remnant
(calyx) of the flower.

Red Ash
Fraxinus pennsylvanica Marsh.

A round-topped tree along streams and in bottomlands. Probably
more common east of the Balcones Escarpment than west of it.
Leaves up to 8 or more inches long, divided into 5 to 9 leaflets
with smooth to slightly toothed margins and pointed tips. Flowers
small, in clusters, male and female on separate trees. Fruits in
conspicuous clusters, dry, winged, resembling a paddle with a
rounded or pointed blade, wing extending alongside the seed
halfway or more to the base.

Texas Ash[*]
Fraxinus texensis (Gray) Sarg.

A small tree of limestone hills and canyons. Leaflets usually 5, rounded, not as elongate as in Red Ash. Wings usually not extending beyond the middle of the seed. (See Red Ash for comparison.)

Japanese Privet *
Ligustrum japonicum Thunb.

Evergreen shrub or small tree. Escaping from cultivation and
established in fencerows, abandoned pastures, and low woodlands.
Twigs greenish brown to gray, without hairs but with raised,
corky dots (lenticels). Leaves opposite, petioled; blade firm
textured, ovate to elliptic, up to 4½ inches long and 2 inches wide,
pointed at the tip, and with smooth margins, upper surface dark
green, smooth, glossy; lower surface lighter with a prominent,
yellow, main vein. Flowers white, about ¼ inch wide, petals bent
back, in broad, dense clusters up to 8 inches long. Fruit berrylike,
dark blue, 5/16 inch long and ¼ inch wide, hanging on into winter.

Chinese Privet
Ligustrum sinense Lour.

Evergreen shrub with spreading branches. An escape from
cultivation, found near streams and in old fencerows. Young twigs
covered with fine hairs visible under a 10x hand lens. Leaves
opposite, with short petioles; blades up to 2 inches long, ovate to
elliptic, usually rounded at the tip, sometimes with a small notch,
tapering to the base, and with smooth margins. Flowers white,
fragrant, about ⅜ inch wide, borne in narrow clusters up to 4
inches long, appearing from March to May. Fruit berrylike, bluish
black, ¼ inch long by ³/₁₆ inch wide, in clusters that bend down
the branchlets bearing them, and hanging on into winter.

Elbow-bush, Stretch-berry *
Forestiera pubescens Nutt.

Shrub with opposite, or nearly opposite, twigs, sometimes on long, arched branches. Common in open woodlands, brushy areas, and near streams. Leaves opposite, with a short petiole; blades up to 1¼ inches long, broadly ovate to narrower, with a rounded or pointed tip, tapering or rounded at the base, and with fine-toothed margins. Flowers not showy, yellow, without petals, borne in small clusters on bare twigs in February and March, male and female on separate bushes. Fruit fleshy, dark blue, with a lighter bloom easily rubbed off, ¼ inch in diameter and slightly longer than wide.

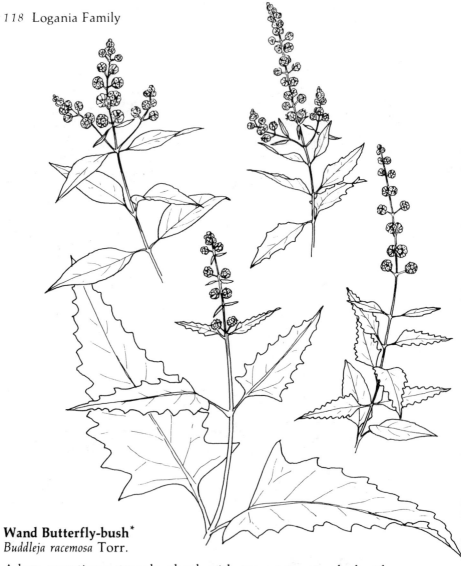

Wand Butterfly-bush[*]
Buddleja racemosa Torr.

A low, sometimes straggly, shrub with gray stems streaked with light brown. Growing from crevices in limestone cliffs and on steep, wooded slopes. Leaf blades variable in size and shape, from ¾ inch long with a wavy margin and a tapered base to 4½ inches long and resembling an arrowhead with deeply scalloped to coarsely toothed edges. Leaf blade upper surface green, the lower grayish green and covered with crystalline glands and fine hairs barely visible under a 10x hand lens. Flowers minute, in tight clusters on shoots extending above the leaves, the clusters and shoots covered with gray or tawny hairs. Flowers present throughout the summer.

Anacua

Ehretia anacua (Teran & Berl.) I.M. Johnst.

Tree with several trunks, or shrub; at the northern limit of its
range in Austin. Planted as an ornamental. Bark gray to reddish
brown, thick, deep furrowed, scaling off into flakes. Leaves
evergreen, some falling seasonally, up to 4½ inches long, mostly
smaller, ovate or narrower, upper surface rough, margins smooth,
tip pointed. Flowers in clusters at the ends of the branchlets,
white, fragrant. Fruit fleshy, spherical, up to 5/16 inch in diameter,
orange to dark yellow, edible.

Texas Lantana, Calico Bush[*]
Lantana horrida H.B.K.

A spreading shrub, much branched from the ground upward, branches sometimes with prickles. Frequent in brushy places and in woodlands. Bark light gray to light brown, tending to flake off. Young twigs nearly square in cross section, covered with short hairs visible under a 10x hand lens. Leaves opposite, up to 2½ inches long, broadly ovate, pointed at the tip, flattened at the base, upper surface rough to the touch; margins coarsely toothed, teeth broad, pointed or rounded. Flowers colorful, red, orange, and yellow, tubular with four flared lobes; in dense, rounded clusters with a leafy bract subtending each flower, at the ends of long, paired stems (peduncles) usually extending beyond the leaves, appearing from April to October. Fruit round, fleshy, dark blue to black.

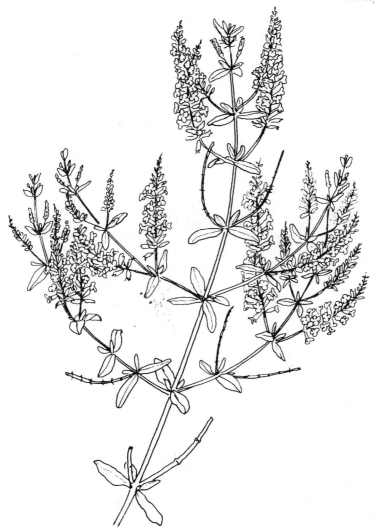

Common Bee-brush, White Brush
Aloysia gratissima (Gill. & Hook.) Troncoso

A fragrant, slender, erect shrub with squarish stems, generally light gray bark, and branches sometimes bearing sharp tips. On rocky upland soils in the Hill Country. Leaves up to 1 inch long by 5/16 inch wide but often smaller, usually in clusters along the stems. Flowers small, white, crowded on spikes up to 3 inches long and extending above the leaves, appearing from March to November.

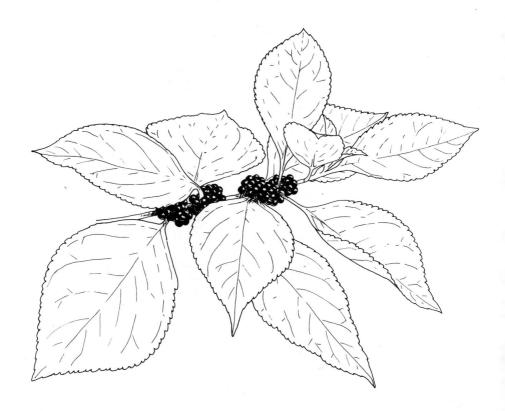

American Beautyberry, French Mulberry*
Callicarpa americana L.

Shrub found on wooded stream terraces, on ledges above streams, and in low, wooded or brushy areas. Bark light brown on the older wood, reddish brown on younger wood. Bark smooth, with elongate, raised corky areas (lenticels); twigs round to 4-sided, covered with branched hairs visible under a 10x hand lens. Leaves in pairs or in threes, blades half as wide as long and up to 9 inches long, ovate to elliptic, pointed or blunt at the tip and tapered to the base; margins coarsely toothed except toward the base and near the tip, teeth pointd or rounded; lower surface of young leaves covered with branched hairs. Flowers small, pink, in dense clusters at the bases of the leaves, clusters usually not exceeding the leaf petioles. Fruit distinctly colored, rose pink or lavender pink, berrylike, about ¼ inch long and 3/16 inch wide, in showy clusters, persisting after the leaves have fallen.

Common Chaste-tree, Sage-tree *
Vitex Agnus-castus L.

A popular landscape shrub occasionally established in the wild.
With smooth gray bark developing ridges and fissures with age.
Young twigs 4-sided, with a soft, velvety surface, older ones
round. Leaves in pairs, consisting of a petiole up to 3 inches long
and 5 to 9 narrow leaflets radiating from its tip; leaflets tapered
from the middle toward each end, the central one usually the
longest, sometimes up to 5 inches; margins smooth to wavy, upper
surface green, lower surface greenish white. Flowers small,
lavender or white, in dense clusters at intervals on a long axis
extending beyond the leaves. Buds densely covered with fine,
white hairs. Fruit round, about ⅛ inch in diameter, dark brown, in
a cup formed by a remnant of the flower.

Tree Tobacco, Mustard Tree
Nicotiana glauca Graham

Tall, narrow, usually evergreen shrub or small tree with green branches whitened by a thin, waxy coating. Established along ditches, roadsides, and streams. Indigenous to Argentina. Leaves up to 7 inches or more long, with a long petiole and ovate to narrower blade pointed at the tip, with smooth margins and a waxy coating on both surfaces. Flowers tubular, yellow with a greenish tinge, 1¼ inches or more long, in loose clusters at the ends of the branches, appearing from spring to winter. Fruit an egg-shaped capsule about ½ inch long, containing many tiny seeds.

Cenizo, Texas Silverleaf
Leucophyllum frutescens (Berl.) I.M. Johnst.

A thickly branched shrub densely covered with stellate hairs on twigs and leaves and having a light gray appearance. Infrequent on well-drained, stony soil, abundant in the South Texas Plains, a popular ornamental. Leaves silvery gray to greenish, soft to the touch, up to 1¼ inches long but mostly 1 inch or less, tapering more gradually to the base than to the rounded tip, margins smooth. Flowers violet to purple, sometimes pink, nearly bell shaped, and up to 1 inch in length and width, appearing intermittently from spring to fall. Fruit a small capsule.

Common Trumpet-creeper, Trumpet-vine
Campis radicans (L.) Seem.

A climbing vine with aerial rootlets. Occasional on low ground,
more common east of the Hill Country. Also cultivated for its
attractive flowers, and escaping cultivation. Leaves with 4 to 6
pairs of leaflets and a terminal one on an axis up to 12 inches long.
Leaflets dark green on the upper surface, lighter on the lower,
broadly to narrowly ovate, with coarse teeth, an elongate tip, and a
rounded to wedge-shaped base, the blade extending along the
petiolule (leaflet stem) to its base. Flowers showy, broadly trumpet
shaped, up to 3½ inches long, orange to reddish orange, clustered
at the ends of branches, appearing throughout the summer. Fruit a
pod up to 6 inches long with 2 ridges running lengthwise, tapering
more gradually to the base than to the tip, and roughly round in
cross section.

Catawba-tree, Northern Catalpa
Catalpa speciosa Warder

Tree with gray to reddish brown scaly bark. More common in East Texas. Planted around homes and occasionally seeding itself near streams. Leaves heart shaped with a drawn-out tip and usually smooth, sometimes shallowly lobed, margins; petioles up to 8 inches long and blades to 12 inches long by 8 inches wide. Flowers showy, up to 2 inches long by 2 inches wide, petals white with yellow streaks and purplish spots inside, fused about half their length and flaring into 2 unequal lips, the smaller 2-lobed, the larger 3-lobed; in open clusters, appearing in May and June. Fruit a long, narrow pod up to 18 inches in length by ½ inch in width.

Common Buttonbush
Cephalanthus occidentalis L.

Shrub with smooth branches. Common on the banks of streams and ponds and in stream beds. Leaves in pairs or in threes, petiolate; blade up to 8 inches long, ovate to narrower, sometimes 1/3 or less as wide as long, with a pointed tip and rounded to tapered base, smooth margins and glossy upper surface, lower surface duller. Flowers small, borne in distinctive, dense, spherical clusters (heads) with a fringe of stamens protruded beyond the white petals, heads 1 to 1½ inches wide. Fruiting structure a head of tightly packed small capsules.

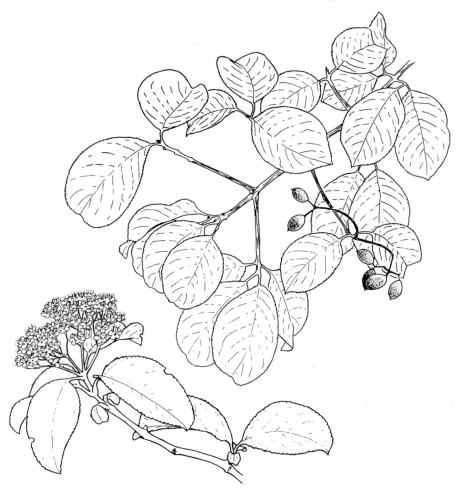

Southern Black-haw, Downy Viburnum, Rusty Nanny-berry*
Viburnum rufidulum Raf.

Tall shrub or small tree with bark separating into dark, rectangular plates. Found on stream banks, in thickets, and in light shade in woodlands. Twigs reddish brown with a thin light gray coating. Leaves in pairs, often on short spurs, the petioles covered with rust-colored, branched hairs visible under a 10x hand lens; blades up to 3½ inches long, elliptic to oval or ovate, tip rounded or with a broad point, base rounded or broadly wedge shaped, margins finely serrate, firm textured with a shiny upper surface. Flowers white, from ¼ to ⅜ inch wide, in rounded or flattened clusters up to 4 inches wide, appearing in March and April and noticeable from a distance in early spring. Fruit fleshy, bluish black lightened by a waxy coating, up to ½ inch long, slightly longer than wide.

Common Elder-berry, American Elder
Sambucus canadensis L.

Tall shrub with a broad, white pith in stems and branches. Found at the edges of streams and in low, wet places. Leaves up to 12 inches long, opposite, consisting of a central axis with 4 to 10, usually 4 to 6, paired leaflets and a terminal one often larger. Leaflets ovate to elliptic or narrower, up to 7 inches long, with an extended tip and a broadly wedge-shaped base; margins toothed except at the tip and toward the base, the teeth narrow and pointed toward the tip. Flowers white, 3/16 to ¼ inch across, in broad, flat, conspicuous clusters up to 10 inches or more in diameter, appearing from May to July. Fruit berrylike, dark purple when ripe, 3/16 to ¼ inch wide, edible.

Japanese Honeysuckle *

Lonicera japonica Thunb.

A common, weedy, twining vine of moist places, shading out shrubs and herbaceous plants and climbing into trees. Leaves green into winter, ovate to oblong, with smooth margins and a short point at the tip, often lobed or toothed on young shoots. Flowers in pairs, fragrant, showy, 1 to 2 inches long, white, turning yellow with age, shaped like narrow tubes flaring into 2 lips, one broader than the other and 4-lobed, appearing in spring and summer. Fruit ¼ inch in diameter, fleshy, black, with a short tip. Naturalized from eastern Asia.

Trumpet Honeysuckle
Lonicera sempervirens L.

A twining, evergreen vine with bicolored foliage, green on the
upper surface and white on the lower. Climbing over shrubbery in
low, moist habitats, at the eastern edge of the Hill Country,
infrequent, more frequent eastward. Leaves ovate to oblong with
smooth, rolled-down margins and a blunt or short-pointed tip,
those immediately below the flowers fused at the base. Flowers
showy, trumpet shaped, up to 2 inches long, mostly bright red,
arranged in whorls at the ends of shoots, appearing in spring. Fruit
a red berry.

White Honeysuckle*
Lonicera albiflora T. & G.

A shrub with occasionally twining branches. Infrequent in cedar-
oak woodlands in the Austin area but apparently frequent to
common elsewhere in the same habitat. Leaves paired, broadly
oval, up to 2¾ inches long by 1¾ inches wide but usually half
these dimensions, rounded at the tip, with a minute point, smooth
margins, and rounded or tapered at the base. The pair of leaves
immediately below the flowers fused at their bases. Flowers white
to yellowish white, two-lipped, up to ⅝ inch long, in clusters at the
ends of the branches, appearing in spring. Fruit a berry up to ⅜
inch in diameter.

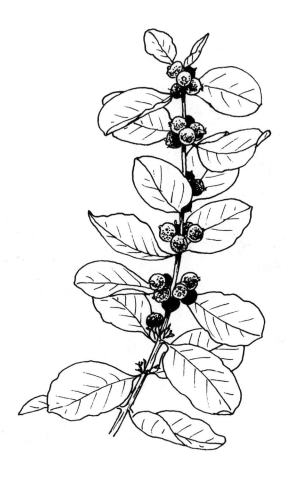

Coral-berry, Indian-currant
Symphoricarpos orbiculatus Moench.

A much-branched shrub. On wooded stream banks at the eastern edge of the Hill Country. With shreddy bark on older wood and brown to purplish branchlets covered with short hairs visible under a 10x hand lens. Branchlets bearing many leaves. Leaves roughly oval, tapering about equally to tip and base, up to 2 inches long but often less than 1 inch, with smooth, turned-down margins and a rounded or broadly pointed tip. Flowers small, 3/16 inch long, in elongate, compact clusters arising from the axils of the leaves, appearing throughout the summer. Fruit fleshy, pink to yellowish red, up to ¼ inch in diameter, persisting into winter.

Shrubby Boneset*
Eupatorium havanense H.B.K.

A shrub with thin stems and smooth gray-to-brown bark, young
twigs greenish or reddish. Frequent to common on rough
limestone slopes. Leaf blades triangular to roughly ovate or
narrower, up to 3 inches long, with 3 main veins; margins wavy to
coarsely toothed, tip pointed, and the base broadly tapered to
almost perpendicular to the petiole. Flowers minute, white, in
heads about 5/16 inch wide and clustered at the ends of branches,
appearing in October and November. Fruit ⅛ inch long, with a
crown of bristles on one end.

Roosevelt Weed, Jara Dulce*
Baccharis neglecta Britt.

A weedy, tall shrub abundant in fields out of cultivation and on disturbed ground, also in unshaded, low places. With ascending light brown branches and green twigs. Leaves partly evergreen, very narrow, less than ¼ inch wide and up to 3 inches long. Male and female flowers on separate plants. Female flowers inconspicuous, silky, in small, greenish white heads which appear to be individual flowers, these arranged in large clusters up to 1 foot or more long and 8 inches wide, resembling silky plumes in October and November. Fruit, minute, about ¹/₁₆ inch long, borne on the wind by a tuft of hairs.

Orange Zexmenia
Zexmenia hispida (H.B.K.) Gray

A shrub up to 3 feet tall, woody at the base and herbaceous in the leafy parts. Common on well-drained limestone slopes and uplands. Upper stems covered with stiff appressed hairs and rough to the touch. Leaves rough on both surfaces, varying in shape from lobed in the basal half on either or both sides to narrow and unlobed, with margins weakly toothed or untoothed. Flowers minute, in showy heads with yellow to orange rays appearing on petals of a single flower, on long stems (peduncles) extending vertically above the leaves, appearing mainly in summer and fall, observed also in mid-May.

Damianita
Chrysactinia mexicana Gray

Low, aromatic, evergreen shrub with small, needlelike leaves and bright yellow flowers. Infrequent on unshaded, stony hillsides and uplands west of the Balcones Escarpment. Stems usually less than 2 feet tall, supporting leafy, upright twigs. Leaves crowded on the twigs, dark green, up to ⅜ inch long by 1/16 inch or less wide, covered with green to black glandular dots visible under a 10x hand lens. Flowers minute, in heads on thin stems (peduncles) up to 3 inches long; heads resembling yellow daisies, appearing from April to early October. Fruit very small, ⅛ inch long and with a crown of short bristles visible under a 10x hand lens.

KEY TO IDENTIFICATION

How to use the Key to Identification

This key consists of a series of paired choices based, for the most part, on contrasting characteristics, i.e. *leaf broad* versus *leaf narrow*. Each pair of choices has a number, and the choices are *a* and *b* under that number. To identify a plant, start with *1a* and match your plant with it. If the plant doesnt't fit, go to *1b*. It will fit one or the other. If it fits *1a*, go next to 2 and decide between *2a* and *2b*. Continue, following the numbers indicated. If your plant fits *1b*, go to 7 (the number indicated) and decide between *7a* and *7b*. The number in parentheses at the end of a choice indicates your next pair of choices. The number in parentheses to the right of a choice number, e.g. *7a* (**1b**), indicates the choice from which you came. Proceed until a choice ends in a plant name. Then turn to the page numbered after it to confirm your identification. The number of choices, or steps, required to key out a plant is usually small. To key out *Agarito, 155a*, the steps are: *1b, 7b, 10b, 19b, 152a, 153a, 154a,* and *155a*.

Some species, because of their variability, appear more than once in the Key. The Hawthorns, because they are extremely difficult to separate at the species level, do not appear. A brief description does occur in the text.

Carefully examine the plant you are trying to identify and avoid using rootsprouts and growth from injured parts. Check the Glossary for terms with which you are not familiar.

(Choice Number)	(Previous Choice Number)	Description	(Next Choice Number) or Plant Name, page reference

1a Leaves minute, scalelike, or from ⅛ to ¾ inch long and ¹/₁₆ inch or less wide (2)
1b Leaves otherwise (7)
2a (1a) Leaves scalelike (3)
2b Leaves not scalelike (5)
3a (2a) Shrub; leaves scarcely noticeable—**Joint Fir**, p. 4
3b Tree, evergreen; leaves overlapping, covering the branchlets (4)
4a (3b) Trunk branching into several trunks at or near the ground line—**Ashe Juniper**, p. 2

4b Trunk single, erect—**Eastern Red Cedar**, p. 3

5a (2b) Leaves up to ¾ inch long (6)

5b Leaves up to ⅜ inch long, dark green, aromatic, low shrub—
Damianita, p.138

6a (5a) Cone-bearing tree—**Bald Cypress**, p. 1

6b Low shrublike plant with thin stems—**Texas Desert-rue**, p. 67

7a (1b) Leaves small, fleshy, falling early in the growing season;
plant spiny with fleshy, green segmented stems (8)

7b Leaves otherwise (10)

8a (7a) Stem segments flattened in cross section (9)

8b Stem segments rounded in cross section—**Pencil Cactus**, p. 105

9a (8a) Stems prostrate, creeping along the ground—**Brownspine
Prickly Pear**, p.106

9b Stems erect—**Texas Prickly Pear**, p.106

10a (7b) Leaves with parallel veins running from the base to the
tip (11)

10b Leaf veins not parallel, branching from a midrib or main veins
(19)

11a (10a) Leaf blades fanlike or palmlike—**Dwarf Palmetto**, p. 6

11b Leaf blades long and narrow (12)

12a (11b) Margins of leaf blades smooth or with very fine teeth
(13)

12b Margins of leaf blades with short, sharp spines—**Sotol**, p.13

13a (12a) Blades with a stiff, sharp point at the tip (14)

13b Blades without a stiff, sharp point (17)

14a (13a) Blades with curly, white fibers on their margins (15)

14b Blades without curly, white fibers on their margins (16)

15a (14a) Leaves numerous, forming a spiny cluster at the ground
line or on a prostrate stem; flowering stalk branched, up to 10
feet tall—**Buckley Yucca**, p. 9

15b Leaves fewer, forming an open cluster at the ground line;
flowering stalk rarely branched and not over 6 feet tall—
Arkansas Yucca, p. 10

16a (14b) Plant stemless, leaves arising from the ground line,
twisted—**Twisted-leaf Yucca**, p.8

16b Shrub or small tree, with a tall stem on old plants; dead leaves
hanging under the live ones—**Spanish Dagger**, p. 7

17a (13b) Leaves forming a clump on the ground, sometimes
resembling a large clump of grass (18)

17b Leaves arranged along a tall stem, a large grass—**Giant Reed,**
p.5

18a (17a) Blades flat in cross section—**Devil's-shoestring**, p.11

18b Blades V-shaped in cross section—**Sacahuista**, p.12

19a (10b) Leaves simple, blade in one piece, not divided into leaflets
(20)

19b Leaves compound, the blades divided into leaflets (152)

20a (19a) Leaves alternately arranged on the branchlets, sometimes
two or more together or on alternate spurs bearing one to

several leaves (21)

20b Leaves opposite or whorled, two or three arising at the same node but on different sides of the stem (126)

21a (20a) Leaves with a petiole, a stem supporting the blade (22)

21b Leaves without a petiole, or with one not more than ¹/₁₆ inch long (120)

22a (21a) Leaf blades not lobed (23)

22b Some or all leaf blades lobed (95)

23a (22a) Blades with smooth to slightly wavy margins (24)

23b Leaf blades with toothed margins (58)

24a (23a) At least some of the leaves on short spurs, spurs generally on older wood (not to be confused with thorn-tipped branchlets) (25)

24b Leaves not on short spurs (27)

25a (24a) Leaves more than 1 inch long (26)

25b Leaves less than 1 inch long, shrub—**Texas Almond**, p. 50

26a (25a) Leaves pointed at the tip, up to 4 inches and more long; tree—**Bois-d'arc**, p. 36

26b Leaves rounded at the tip; tree—**Coma**, p. 110

27a (24b) Leaf blades narrowing to a point at the tip (28)

27b Leaf blades rounded to indented at the tip, sometimes with an abrupt, short point (51)

28a (27a) Leaf blades tapered to the base, the margins of the blades more or less straight and meeting the petiole at an angle (29)

28b Leaf blades not tapered to the base (36)

29a (28a) Trees (30)

29b Shrubs (31)

30a (29a) Blades about as broad as long—**Chinese Tallow Tree**, p. 77

30b Blades up to three times as long as broad—**Laurel Cherry**, p. 48

31a (29b) Leaves more than 2 inches long (32)

31b Leaves not more than 2 inches long (35)

32a (31a) Branches green—**Tree Tobacco**, p. 124

32b Branches not green (33)

33a (32b) Veins prominent on the lower surface of the leaf—**Indian-cherry**, p. 90

33b Veins not prominent on the lower surface of the leaf (34)

34a (33b) Leaf blades about as broad as long—**Sycamore-leaf Snow-bell**, p. 112

34b Leaf blades about twice as long as broad—**Spicebush**, p. 43

35a (31b) Leaves bright green, borne on thin but stiff, thorn-tipped branchlets—**Brasil**, p. 94

35b Leaves not bright green, borne on rather thick, stout, thorn-tipped branchlets—**Lotebush**, p. 93

36a (28b) Leaf blades rounded at the base (37)

36b Leaf blade lobed to heart shaped at the base, or more or less flat across the base, truncate (47)

37a (36a) Widest part of the blade nearer the base than the tip (38)

37b Widest part of the blade at the middle or toward the tip, or the margins nearly parallel (43)

38a (37a) Trees (39)

38b Shrubs (42)

39a (38a) Blades about as broad as long—**Chinese Tallow Tree**, p. 77

39b Blades two to three times as long as broad (40)

40a (39b) Blades asymmetric at the base, one side more rounded than the other (41)

40b Blades not asymmetric at the base, bark on trunk and limbs smooth, pink—**Texas Madrone**, p. 109

41a (40a) Leaves up to 4 inches long—**Sugar Hackberry**, p. 29

41b Leaves not over 3 inches long—**Netleaf Hackberry**, p. 30

42a (38b) Leaves up to 7 inches long, smooth on both surfaces, branches green—**Tree Tobacco**, p. 124

42b Leaves not over 3 inches long, velvety on the lower surface— **Encinilla**, p. 75

43a (37b) Trees (44)

43b Not Trees (46)

44a (43a) Bark on the trunk and limbs smooth, pink—**Texas Madrone**, p. 109

44b Bark on trunk rough (45)

45a (44b) Upper surface of the leaves rough to the touch—**Anacua**, p. 119

45b Leaves smooth—**Plateau Live Oak**, p. 25

46a (43b) Shrub; veins prominent on the lower surface of the leaf —**Indian-cherry**, p. 90

46b Vine, stems green—**Alabama Supple-jack**, p. 91

47a (36b) Shrubs or small trees (48)

47b Vines (50)

48a (47a) Leaves broadly heart shaped, rose to purplish flowers appearing in early spring—**Redbud**, p. 58

48b Leaves and flowers otherwise (49)

49a (48b) Leaves broadly ovate to almost circular, up to 4 inches long; shrub—**Sycamore-leaf Snow-bell**, p. 112

49b Leaves ovate to narrower than ovate, less than 3 inches long, velvety on the lower surface, aromatic—**Encinilla**, p. 75

50a (47b) Stems bearing sharp prickles—**Cat-brier**, p. 14

50b Stems without prickles—**Carolina Snailseed**, p. 42

51a (27b) Leaf blades tapered to the base, the margins of the blades more or less straight and meeting the petiole at an angle (52)

51b Leaf blades not tapered to the base (55)

52a (51a) Most leaves more than 1½ inches long (53)

52b Most leaves less than 1½ inches long; shrubs (54)

53a (52a) Veins prominent on the lower surface of the blade; shrub or small tree—**Indian-cherry**, p. 90

53b Veins not prominent on the lower surface of the blade; shrub to medium-sized tree; small thorns among the leaves—**Coma**, p. 110

54a (52b) Leaves bright green, borne on thin but stiff thorn-tipped branchlets—**Brasil**, p. 94

54b Leaves not bright green, borne on rather thick, stout, thorn-tipped branchlets—**Lotebush**, p. 93

55a (51b) Leaf blades rounded at the base, prominent veins branching from a midrib, stems green; a vine in canyons—**Alabama Supple-jack**, p. 91

55b Leaf blades flat, truncate, or lobed to heart shaped at the base (56)

56a (55b) Shrub or small tree—**Redbud**, p. 58

56b Vine (57)

57a (56b) Stems green, bearing sharp prickles—**Cat-brier**, p. 14

57b Stems without prickles—**Carolina Snailseed**, p. 42

58a (23b) At least some of the leaves on short spurs, very short in 61b (59)

58b Leaves not on short spurs (62)

59a (58a) Teeth on leaf margins rounded, blades tapering more gradually to the base than to the tip; tall shrub or small tree—**Possumhaw**, p. 82

59b Teeth pointed (60)

60a (59b) Leaves more than 1¼ inches long—**Creek Plum**, p. 52

60b Leaves less than 1¼ inches long (61)

61a (60b) Leaves dark green on the upper surface, grayish green and densely hairy on the lower surface, blades often twisted, the margins rolled down—**Brush Myrtle-Croton**, p. 76

61b Leaves with fine hairs, but not densely hairy on the lower surface, blades not twisted—**Texas Colubrina**, p. 88

62a (58b) Leaf blades narrowing to a point at the tip (63)

62b Leaf blades rounded or indented at the tip, sometimes with an abrupt, short point (92)

63a (62a) Leaf blades tapered to the base and meeting the petiole at an angle, the margins more or less straight (64)

63b Leaf blades not tapered to the base (71)

64a (63a) Widest part of the blade nearer the base than the tip, blade long and narrow—**Black Willow**, p. 15

64b Leaf blades otherwise (65)

65a (64b) Widest part of the blade nearer the tip than the base (66)

65b Widest part of the blade in the middle or the margins nearly parallel (69)

66a (65a) Shrub with stiff, thorn-tipped branchlets (67)

66b Branchlets not thorn tipped (68)

67a (66a) Thorn-tipped branchlets thin, leaves bright green—**Brasil**, p. 94

67b Thorn-tipped branchlets rather thick, stout, leaves not bright green—**Lotebush**, p. 93

68a (66b) Leaves up to 8 inches long; tree—**Chinkapin Oak**, p. 22

68b Leaves less than 5 inches long; immature tree—**Laurel Cherry**, p. 48

69a (65b) Veins on the lower side of the leaf yellowish and prominent (70)

69b Veins on the lower side of the leaf not prominent, blade usually more than 3 inches long, shiny on the upper surface; immature tree—**Laurel Cherry**, p. 48

70a (69a) Shrub or small tree with leaves up to 5 inches long, teeth minute—**Indian-cherry**, p. 90

70b Shrub with leaves less than 3 inches long—**Redroot**, p. 89

71a (63b) Blades rounded at the base (72)

71b Blades otherwise (84)

72a (71a) Shrubs (73)

72b Trees (76)

73a (72a) Leaves up to 5 inches long, teeth minute, veins on the lower side of the leaf yellowish, prominent—**Indian-cherry**, p. 90

73b Leaves less than 3 inches long (74)

74a (73b) Veins on the lower side of the leaf yellowish, prominent —**Redroot**, p. 89

74b Veins on the lower side of the leaf not prominent (75)

75a (74b) Teeth on the leaf margins easily visible to the naked eye —**Wright Pavonia**, p. 104

75b Teeth not easily visible, visible under a 10x hand lens, leaf surfaces like felt—**Encinilla**, p. 75

76a (72b) Leaves firm textured, dark green and lustrous on the upper surface, light green on the lower surface, teeth pointed, few teeth per leaf except on rootsprouts; bark gray to black, rough—**Plateau Live Oak**, p. 25

76b Leaves otherwise (77)

77a (76b) Lower surface of the leaves with a prominent midrib and two lateral veins arising from the base of the blade, blade less than 3 inches long—**Jujube**, p. 92

77b Leaves otherwise (78)

78a (77b) Leaf blades usually asymmetric, oblique, at the base (79)

78b Leaves not asymmetric, oblique at the base (82)

79a (78a) Bark gray, warty, or with corky ridges, not scaly (80)

79b Bark gray to light brown, scaly (81)

80a (79a) Leaves up to 4 inches long—**Sugar Hackberry**, p. 29

80b Leaves less than 3 inches long—**Netleaf Hackberry**, p. 30

81a (79b) Leaves up to 6 inches long, tapering to a narrow tip— **American Elm**, p. 32

81b Leaves less than 2½ inches long, with a blunt tip, twigs on young plants often with winged bark—**Cedar Elm**, p. 31

82a (78b) Sap milky, upper surface of the leaf rough to the touch— **Paper Mulberry**, p. 37

82b Sap not milky; bark, where not roughened, with horizontal lines (83)

83a (82b) Blades shiny on the upper surface—**Black Cherry**, p. 49

83b Blades dull on the upper surface—**Mexican Plum**, p 51

84a (71b) Blades lobed to heart shaped at the base (85)

84b Blades truncate, more or less flat at the base (89)

85a (84a) Shrub or small tree (86)

85b Vine resembling a grapevine—**Heart-leaf Ampelopsis**, p. 99

86a (85a) Bark gray, warty, or with corky ridges (87)

86b Bark not warty or with corky ridges (88)

87a (86a) Leaves up to 4 inches long—**Sugar Hackberry**, p. 29

87b Leaves less than 3 inches long—**Netleaf Hackberry**, p. 30

88a (86b) Leaf surfaces like felt, covered with stellate hairs visible under a 10x hand lens; shrub—**Encinilla**, p. 75

88b Leaf surface not covered with stellate hairs—**Texas Mulberry**, p. 33

89a (84b) Margins of the blade at the base perpendicular to the petiole or slightly lobed (90)

89b Margins of the blade oblique at the base, meeting the petiole at an angle—**Carolina Basswood**, p. 102

90a (89a) Tree, attaining a large size—**Eastern Cottonwood**, p. 17

90b Not a tree (91)

91a Shrub—**Wright Pavonia**, p. 104

91b Vine resembling a grapevine—**Heart-leaf Ampelopsis**, p. 99

92a (62b) Leaf blades tapered to the base, the margins of the blades more or less straight and meeting the petiole at an angle (93)

92b Leaf blades rounded at the base (94)

93a (92a) Leaves bright green, borne on thin but stiff thorn-tipped branchlets—**Brasil**, p. 94

93b Leaves not bright green, borne on rather thick, stout, thorn-tipped branchlets—**Lotebush**, p. 93

94a (92b) Teeth pointed, few or none per leaf except on rootsprouts—**Plateau Live Oak**, p. 25

94b Teeth rounded, small, covering most of the blade margin—**Yaupon**, p. 83

95a (22b) Lobes with teeth on their margins, teeth not restricted to the ends of the lobes (96)

95b Lobes without teeth on their margins, teeth restricted to the ends of the lobes (109)

96a (95a) Blades deeply trilobed or parted; vine with fleshy leaves —**Ivy Treebine**, p. 101

96b Plant and blades not as above (97)

97a (96b) Trees (98)

97b Not Trees (103)

98a (97a) Leaves bilaterally symmetric, both halves of the blade similarly lobed (99)

98b Lobed leaves not bilaterally symmetric, lobes often dissimilar on the two sides of the blade, or one side lobed and the other unlobed (100)

99a (98a) Lobes with long, narrow tips, bark smooth, light colored, peeling off in thin plates—**Sycamore**, p. 45

99b Lobes with broad tips, lower surface of the blades white— **White Poplar**, p. 16

100a (98b) Blades less than 2½ inches long—**Texas Mulberry**, p. 33

100b Blades more than 2½ inches long (101)

101a (100b) Petioles up to 4 inches long, leaves mostly unlobed, rough on the upper surface, velvety on the lower surface—**Paper Mulberry**, p. 37

101b Petioles usually not more than 2 inches long, blades often deeply lobed (102)

102a (101b) Blades velvety on the lower surface—**Red Mulberry**, p. 35

102b Blades not velvety on the lower surface, smooth to slightly rough—**White Mulberry**, p. 34

103a (97b) Shrubs (104)

103b Woody vines, grapes (107)

104a (103a) Lobes broad, shallow (105)

104b Leaf blades unlobed to deeply lobed (106)

105a (104a) Leaf blades up to 3½ inches long and equally wide, flowers bright red—**Drummond Wax-mallow**, p. 103

105b Leaf blades usually not more than 2 inches long, flowers rose colored—**Wright Pavonia**, p. 104

106a (104b) Leaves less than 2½ inches long—**Texas Mulberry**, p. 33

106b Leaves up to 8 inches long, broad, deeply lobed, twigs thick, sap milky—**Common Fig**, p. 38

107a (103b) Leaves velvety and white on the lower surface— **Mustang Grape**, p. 95

107b Leaves green on the lower surface (108)

108a (107b) Flowers and fruits in clusters up to 8 inches long— **Spanish Grape**, p. 97

108b Flowers and fruits in clusters usually not more than 3 inches long—**Sweet Mountain Grape**, p. 96

109a (95b) Blades with two broad lateral lobes below the apex (110)

109b Blades otherwise (111)

110a (109a) Tree, widest part of the blade toward the apex, lobes with bristlelike tips ¹/₁₆ inch long—**Blackjack Oak**, p. 28

110b Shrub; leaves broad—**Sycamore-leaf Snow-bell**, p. 112

111a (109b) Blades pinnately lobed (112)

111b Blades palmately lobed (118)

112a (111a) Lobes with short, pointed tips, or bristlelike tips (113)

112b Lobes without short, pointed tips or bristles (116)

113a (112a) Lobes separated by deep sinuses, tips up to ³/₁₆ inch long, fruit an acorn (114)

113b Lobes not separated by deep sinuses, fruit an acorn (115)

114a (113a) A tree common on limestone hills—**Texas Oak**, p. 27

114b A tree of canyons and narrow valleys—**Shumard Oak**, p. 26

115a (113b) Lobes many, small; blade margins almost wavy, or margins coarsely toothed, bristlelike tips on lobes less than 1/16 inch long—**Chinkapin Oak**, p. 22

115b Leaves mostly three lobed, bristlelike tips 1/16 inch long— **Blackjack Oak**, p. 28

116a (112b) Trees; lobes separated by deep sinuses, fruit an acorn (117)

116b Small tree or shrub; leaf margin wavy to lobed, fruit an acorn, found on flat-topped limestone hills—**White Shin Oak**, p. 24

117a (116a) Leaves up to 9 inches long, acorn cup with coarse scales and a fringed margin; found along streams—**Bur Oak**, p. 21

117b Leaves 3 to 5 inches long or longer; found on sandy and gravelly soils— **Post Oak**, p. 23

118a (111b) Lobes pointed; tree with smooth, light-colored bark; found near streams—**Sycamore**, p. 45

118b Lobes rounded; vines (119)

119a (118b) Stems bearing sharp prickles—**Cat-brier**, p. 14

119b Stems without prickles—**Carolina Snailseed**, p. 42

120a (21b) Leaf blades with smooth margins (121)

120b Leaves with toothed margins (125)

121a (120a) Leaves silvery gray or greenish and covered with fine hairs on one or both surfaces (122)

121b Leaves green (123)

122a (121a) Lower surface of leaves silvery gray or greenish and covered with fine hairs, branchlets spine tipped and bearing stout lateral spines—**Allthorn Goatbush**, p. 71

122b Both surfaces of leaves silvery gray or greenish and covered with fine hairs, branchlets not spiny—**Cenizo**, p. 125

123a (121b) Leaves not more than three times as long as wide (124)

123b Leaves long and narrow, up to ten or more times as long as wide; shrub—**Roosevelt Weed**, p. 136

124a (123a) Shrub or small tree with smooth, light gray bark— **Texas Persimmon**, p. 111

124b Low shrub with thin stems, usually less than 3 feet tall— **Maidenbush**, p. 74

125a (120b) Leaves about twice as long as broad, often appearing twisted or distorted, dark green on the upper surface, grayish green and densely hairy on the lower surface, the margins rolled down—**Brush Myrtle Croton**, p. 76

125b Leaves long and narrow, up to ten or more times as long as wide, sparsely toothed—**Roosevelt Weed**, p. 136

126a (20b) Leaf blades supported by a petiole (127)

126b Leaves without a petiole, or petiole not more than $1/16$ inch long (147)

127a (126a) Leaf blades with smooth margins (individuals of Wand Butterfly-bush and Shrubby Boneset occasionally with smooth margins but treated under 127b) (128)

127b Leaf blades with toothed margins (140)

128a (127a) Leaf blades narrowing to a point at the tip (129)

128b Leaf blades rounded to indented at the tip, sometimes with an abrupt, short point (136)

129a (128a) Blades tapered or rounded at the base (130)

129b Blades truncate, flat or lobed to heart shaped at the base; tree with large leaves—**Catawba-tree**, p.127

130a (129a) Shrub or small tree (131)

130b Vine climbing over shrubbery and into trees in low places—**Japanese Honeysuckle**, p.131

131a (130a) Most of the blades extended into a narrow tip (132)

131b Blades not extended into a narrow tip (134)

132a (131a) Leaves opposite or whorled; a shrub at the edges of streams and ponds—**Common Buttonbush**, p.128

132b Leaves opposite only (133)

133a (132b) Upper surface of the blade smooth, glossy—**Japanese Privet**, p.115

133b Upper surface of the leaf somewhat rough, not glossy—**Rough-leaf Dogwood**, p.108

134a (131b) Blades leathery, with a short tip $1/16$ inch long—**Lindheimer Silk-tassel**, p.107

134b Blades not leathery (135)

135a (134b) Blades with three prominent veins on the upper surface—**Mock Orange**, p.44

135b Blades with one main vein (midrib)—**Coral-berry**, p.134

136a (128b) Blades thick, leathery (137)

136b Blades not thick or leathery (138)

137a (136a) A parasite on the branches of trees—**Mistletoe**, p. 39

137b Shrub, not parasitic on trees, leaves smooth on the upper surface, velvety on the lower—**Lindheimer Silk-tassel**, p. 107

138a (136b) Shrubs (139)

138b Vine climbing over shrubbery and into trees in low places—**Japanese Honeysuckle**, p. 131

139a (138a) Bark on main branches loose, shreddy—**Coral-berry**, p.134

139b Bark on main branches tight—**Chinese Privet**, p.116

140a (127b) Tree with milky sap—**Paper Mulberry**, p.37

140b Sap not milky (141)

141a (140b) Shrub or small tree (142)

141b Vine; leaves toothed on new growth—**Japanese Honeysuckle**, p.131

142a (141a) Blade margins very finely toothed, blades mostly less than $1\frac{1}{2}$ inches long when fully expanded—**Elbow-bush**, p.117

142b Teeth easily visible to the naked eye, fully expanded blades more than 1½ inches long (143)

143a (142b) Teeth small, pointed, about seven per ½ inch on fully extended leaves; tall shrub, or small tree—**Southern Black-haw**, p.129

143b Teeth fewer than seven per ½ inch (144)

144a (143b) Leaves ovate to elliptic, many more than 5 inches long, the blade tapering to the petiole—**American Beautyberry**, p.122

144b Leaves typically less than 5 inches long (145)

145a (144b) Lower surface of the leaf grayish green covered with crystalline glands visible under a 10x hand lens, common on limestone cliffs—**Wand Butterfly-bush**, p.118

145b Lower surface of the leaf green (146)

146a (145b) Upper surface of the leaf rough to the touch—**Texas Lantana**, p.120

146b Upper surface of the leaf smooth—**Shrubby Boneset**, p.135

147a (126b) Leaf blades with smooth to slightly wavy margins (148)

147b Leaf blades with lobed or toothed margins, stem woody at the base, herbaceous above, up to 3 feet tall—**Orange Zexmenia**, p.137

148a (141a) Leaves broadly oval to elliptic or oblong, from 1 to 2½ inches long (149)

148b Leaves not broadly oval to elliptic, small, or long and narrow (151)

149a (148a) Leaves thick, leathery; a parasite on trees—**Mistletoe**, p.39

149b Leaves not thick or leathery (150)

150a (149b) Leaves green on the upper surface, white on the lower surface; a twining vine—**Trumpet Honeysuckle**, p.132

150b Leaves green on both surfaces; a shrub with occasionally twining branches—**White Honeysuckle**, p.133

151a (148b) Leaves usually less than 1 inch long and 5/16 inch wide, small leaves often clustered with larger ones—**Common Bee-brush**, p.121

151b Leaves narrow, more than 1 inch long and up to ½ inch wide, stem herbaceous except at the base, seldom over 18 inches tall—**Narrow-leafed Thryallis**, p.73

152a (19b) Leaves alternately arranged on the branchlets, sometimes two or more arising from the same place (153)

152b Leaves opposite or whorled, two or three arising from the same node but on different sides of the stem (191)

153a (152a) Leaves trifoliolate, the blades divided into three leaflets (154)

153b Leaves not trifoliolate (160)

154a (153a) Shrubs or small trees (155)

154b Vines (159)

155a (154a) Leaflets stiff, spine tipped—**Agarito**, p.40

155b Leaflets otherwise (156)

156a (155b) Stems trailing on the ground, covered with prickles—**Southern Dewberry**, p 46

156b Stems not both trailing and prickly (157)

157a (156b) Leaflets toothed on their margins (158)

157b Leaflet margins smooth or with very small, rounded teeth, crushed leaves producing a fetid odor; shrub or small tree—**Wafer-ash**, p. 68

158a (157a) Teeth rounded or blunt and toward the tip of the leaflet, ½ to ⅔ of the leaflet blade with smooth margins tapered to the base; shrub—**Fragrant Sumac**, p. 81

158b Leaflets lobed and toothed, the terminal one three lobed, the laterals usually two lobed (one side of the blade unlobed)—**Poison Ivy**, p. 78

159a (154b) Leaves thick, fleshy, emitting a fetid odor when crushed, stems bearing tendrils—**Ivy Treebine**, p. 101

159b Leaves not thick and fleshy, stems not tendril bearing, but often fastened to trees by aerial rootlets—**Poison Ivy**, p. 78

160a (153b) Leaves pinnately compound, divided into two or more pairs of leaflets or pinnae, one pair in Honey Mesquite, along a common axis (161)

160b Leaves palmately compound or approximately so, leaflets radiating from the tip of the petiole (189)

161a (160a) Leaves once compound, the axis unbranched (162)

161b Leaves twice compound, the main axis branched into leaflet-bearing branches, pinnae (182)

162a (161a) Leaflets with smooth margins (163)

162b Leaflets with toothed margins (173)

163a (162a) Trees (164)

163b Shrubs (166)

164a (163a) Leaflets with an elongate, pointed tip, the terminal leaflet often absent—**Western Soap-berry**, p. 86

164b Leaflets with a blunt, rounded, or indented tip, sometimes with a short point less than ¹⁄₁₆ inch long (165)

165a (164b) Branchlets bearing short spines—**Black Locust**, p. 66

165b Branchlets not spiny—**Eve's Necklace**, p. 61

166a (163b) Leaflets numerous, small, ½ inch or less long (167)

166b Leaflets more than ½ inch long (168)

167a (166a) Leaves up to 3½ inches long and leaflets to ⅜ inch long, releasing a fetid odor when crushed—**Texas Kidneywood**, p. 63

167b Leaves up to 1 inch long and leaflets to ⁵⁄₁₆ inch long—**Black Dalea**, p. 64

168a (166b) Leaflets firm textured, shiny on the upper surface, evergreen (169)

168b Leaflets not firm textured and evergreen (170)

169a (168a) Leaflets rounded or indented at the tip, fruit a hard pod—**Texas Mountain Laurel**, p. 60

169b Leaflets narrowed to a blunt or pointed, but not indented, tip; fruits small, red, in clusters—**Evergreen Sumac**, p. 80

170a (168b) Leaf axis winged with blade tissue between the leaflets —**Prairie Flameleaf Sumac**, p. 79

170b Leaf axis not winged between the leaflets (171)

171a (170b) Leaflets narrow, ¼ inch or less wide and up to 1½ inches long, fruit a winged pod—**Rattlebush**, p. 65

171b Leaflets ¼ inch or more wide (172)

172a (171b) Lower surface of the leaflets covered with fine hairs and often speckled with glandular dots visible under a 10x hand lens, flowers and fruits crowded in spikelike clusters—**Indigo Bush**, p. 62

172b Lower surface of leaflets not gland dotted, flowers and fruits not in spikelike clusters—**Eve's Necklace**, p. 61

173a (162b) Trees (174)

173b Shrubs (178)

174a (173a) Leaflet margins with a few teeth near the base, fruit winged—**Tree-of-Heaven**, p. 70

174b One or both leaflet margins toothed most of their length (175)

175a (174b) Fruit an elongate nut with a smooth shell, in a green husk; leaflets sickle shaped, bark scaly, light grayish brown— **Pecan**, p. 20

175b Fruit a somewhat flattened spherical nut with a rough shell, in a green husk, bark dark brown to dark gray or black (176)

176a (175b) Leaves up to 2 feet long, leaflets up to 5 inches long by 2 inches wide, generally east of the Hill Country—**Black Walnut**, p. 18

176b Leaves not over 14 inches long, in canyons and along streams in the Hill Country (177)

177a (176b) Trunk usually single, up to 3 feet thick—**Arizona Walnut**, p. 18

177b Trunks often several and branching near the ground, often a large shrub—**Little Walnut**, p. 19

178a (173b) Leaflets firm textured, stiff, with spine-tipped teeth— **Texas Barberry**, p. 41

178b Leaflets otherwise (179)

179a (178b) Branchlets bearing prickles (180)

179b Branchlets not prickly—**Mexican Buckeye**, p. 87

180a (179a) Main axis of the leaf with prickles (181)

180b Main axis of the leaf without prickles—**Macartney Rose**, p. 47

181a (180a) Erect shrubs up to 10 feet tall, leaves dark green and shiny on the upper surface—**Prickly Ash**, p. 69

181b Low, trailing shrub—**Southern Dewberry**, p. 46

182a (161b) Leaflets with smooth margins (183)

182b Leaflets with toothed margins (188)

183a (182a) Leaflets more than 1 inch long, main axis usually with one pair of pinnae, branchlets with spines up to 2 inches long;

shrub or small tree—**Honey Mesquite**, p. 55

183b Leaflets less than 1 inch long (184)

184a (183b) Bark green on the main branches, leaflets on long, flat, green axes, spines at the bases of the leaves—**Retama**, p. 59

184b Bark not green (185)

185a (184b) Leaflets up to ½ inch long; shrub or small tree; branchlets bearing prickles—**Catclaw Acacia**, p. 54

185b Leaflets less than ½ inch long (186)

186a (185b) Branchlets with sharp, recurved prickles; shrub (187)

186b Two short, straight spines at the base of the fully expanded leaf, leaflets usually up to 3/16 inch long, as many as twenty-five pairs per pinna; shrub or small tree—**Huisache**, p. 53

187a (186a) One to three pairs of pinnae, leaflet-bearing branches, from the main leaf axis, leaflets up to ⅜ inch long—**Pink Mimosa**, p. 56

187b Four to eight pairs of pinnae from the main leaf axis, leaflets up to 3/16 inch long—**Cat's-claw Mimosa**, p. 56

188a (182b) Fast-growing tree, common in and around towns and cities—**Chinaberry**, p. 72

188b Tendril-bearing vine of stream bottoms and wooded canyons —**Pepper-vine**, p. 100

189a (160b) Tendril-bearing vines (190)

189b Low, trailing prickly shrub—**Southern Dewberry**, p. 46

190a (189a) Usually five leaflets, up to 6 inches long—**Virginia Creeper**, p. 98

190b Usually seven leaflets, up to 2½ inches long—**Seven-leaf Creeper**, p. 98

191a (152b) Leaves trifoliolate, the blades divided into three leaflets, tree—**Boxelder**, p. 84

191b Leaves not trifoliolate (192)

192a (191b) Leaves pinnately compound, leaflets five or more along a common axis (193)

192b Leaves palmately compound, leaflets radiating from the tip of the petiole (197)

193a (192a) Trees (194)

193b Not Trees (196)

194a (193a) Leaflets irregularly lobed and toothed—**Boxelder**, p. 84

194b Leaflets not lobed (195)

195a (194b) Leaflets up to 3 inches long and 2 inches wide, rounded to abruptly pointed at the tip—**Texas Ash**, p. 114

195b Leaflets up to 6 inches long and 2 inches wide, narrowing gradually to the tip—**Red Ash**, p. 113

196a (193b) Shrub near streams and in low, wet places—**Common Elder-berry**, p. 130

196b Vine climbing over shrubs and into trees—**Common Trumpet-creeper**, p. 126

197a (192b) Leaflets grayish green to white on the lower surface, margins smooth to wavy; shrub—**Common Chaste-tree**, p. 123
197b Leaflets green on the lower surface, margins finely toothed; shrub or small tree—**Red Buckeye**, p. 85

ILLUSTRATED GLOSSARY

LEAF ARRANGEMENT

Alternate

Opposite

Whorled

LEAF PARTS

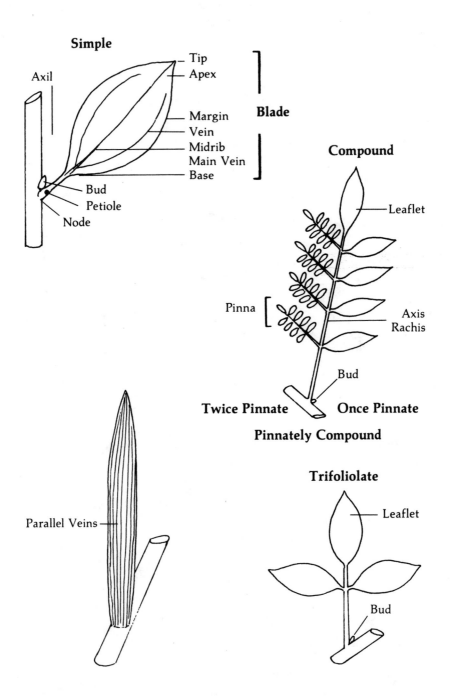

Simple

Axil

Tip
Apex

Margin
Vein
Midrib
Main Vein
Base

Blade

Bud
Petiole
Node

Compound

Leaflet

Pinna

Axis
Rachis

Bud

Twice Pinnate　**Once Pinnate**

Pinnately Compound

Parallel Veins

Trifoliolate

Leaflet

Bud

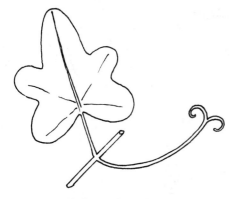

Palmately Lobed

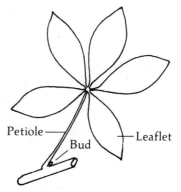

Petiole

Bud

—Leaflet

Palmately Compound

LEAF SHAPES

Ovate

Obovate

Elliptic

Oblong

Heart Shaped

Triangular

LEAF MARGINS

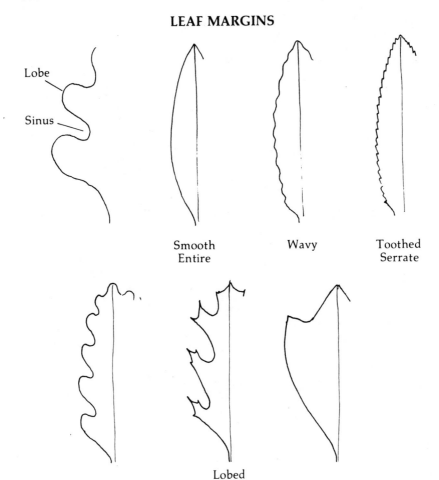

Lobe

Sinus

Smooth
Entire

Wavy

Toothed
Serrate

Lobed

LEAF BASES

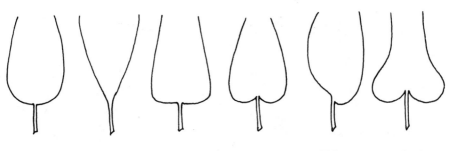

Round   Tapered   Truncate   Heart   Oblique   Lobed
                                    Shaped   Asymmetric

LEAF SURFACES

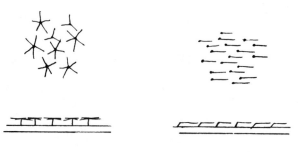

Stellate Hairs   Appressed Hairs

STEM PARTS

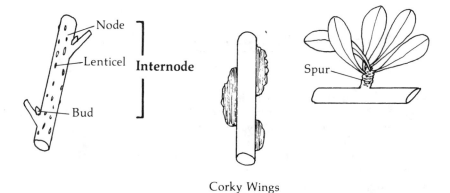

Node
Lenticel   Internode
Bud

Corky Wings

Spur

FLOWERING STRUCTURES

Peduncle

Ray

Peduncle

Head
Many Minute Flowers

FRUIT TYPES

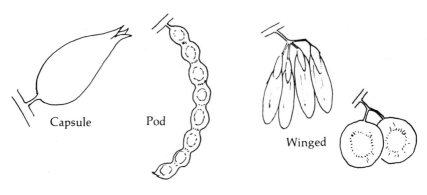

Capsule   Pod   Winged

GLOSSARY

Apical. situated at or near the tip.

Appressed. pressed or lying flat against something.

Asymmetric. lacking symmetry, e.g., the two halves of a leaf blade not alike, resulting in an asymmetric leaf.

Axil. the upper angle between a leaf and the branch or branchlet supporting it.

Axis. a longitudinal support along which parts are arranged, the main stem.

Base. the part of a leaf toward the branch which supports it.

Blade. the broad or expanded part of a leaf.

Bract. a modified leaf below a flower, usually occuring with a cluster of flowers.

Branchlet. a small branch.

Bud. a protuberance on a stem containing rudimentary foliage or floral parts, or both.

Capsule. a dry fruit that splits open along three or more lines.

Compound. divided into parts, e.g., a leaf that is divided into leaflets.

Elliptic. shaped like an ellipse, resembling a flattened circle.

Head. a compact cluster of flowers attached to essentially the same point on the penducle.

Herbaceous. of the texture and color of leaves, non-woody.

Internode. the section of a stem between two nodes.

Leaflet. a unit of a compound leaf.

Lenticel. a corky pore in young bark.

Lobe. a rounded or pointed extension of an organ, e.g., a lobe on a leaf.

Midrib. the central vein or rib of a leaf.

Node. the point on a stem at which a leaf arises.

Oblong. much longer than broad with sides nearly parallel.

Obovate. inversely ovate, with the narrower end of the leaf blade toward the branch.

Ovate. egg shaped, with the broader end of the leaf blade toward the branch.

Palmate. like a spread hand, e.g., leaf parts radiating from the tip of the petiole.

Pedicel. a stalk supporting a single flower in a cluster of flowers.

Peduncle. the stalk supporting a cluster of flowers or a single flower by itself.

Pendulous. suspended, hanging.

Petiole. a leaf stalk, the stalk supporting a leaf blade.

Petiolule. a leaflet stalk.

Pinna (pinnae, pl.). a division of a pinnately compound leaf.

Pinnate. said of a compound leaf with the leaflets arranged along a central axis.

Pistil. the female part of a flower.

Pith. spongy tissue in the center of a stem.

Pod. a dry fruit that splits after ripening, a term applied to fruits in the Legume Family.

Prickle. a sharp process on the surface of a twig or leaf, an outgrowth.

Rachis. the axis of a pinnately compound leaf.

Ray. the fused petals of a minute flower in the margin of a head in the Sunflower Family.

Serrate. toothed like a saw.

Simple. not divided into parts, e.g., a leaf with the blade in one piece.

Sinus. the depression or recess between two lobes.

Spine. a sharp-pointed structure commonly related to a leaf in origin.

Spur. a short, compact twig with very closely spaced nodes.

Stamen. the male part of the flower.

Stellate. star shaped, said of certain branched hairs.

Stock. the main stem of the plant.

Tendril. a slender coiling or twining structure modified from a leaf, branch, or other organ.

Thorn. a sharp-pointed structure formed by a modified branch.

Trifoliolate. having three leaflets.

Truncate. ending abruptly, e.g., a leaf blade squared at the base.

Twig. a small branch.

Vein. a threadlike conduction and support structure in a leaf.

Whorled. arranged in a circle.

Wing. a thin extension of a plant part, as of a branch or dry fruit.

SELECTED REFERENCES

Burlage, H. M. 1973. *The Wild Flowering Plants of Highland Lakes Country of Texas*. Publ. by the author, 158 pp.

Correll, D. S. and M. C. Johnston. 1970. *Manual of the Vascular Plants of Texas*. Texas Research Foundation, Renner, Texas, 1881 pp.

Gould, F. W. 1975. *Texas Plants—A Checklist and Ecological Summary*. Texas A. & M. Exp. Sta. Misc. Pub. 585/Revised, 121 pp.

Kutac, E. A. and S. C. Caran. 1976. *A Bird Finding and Naturalist's Guide for the Austin, Texas, Area*. Travis Audubon Society and Austin Natural Science Association, 145 pp.

Lynch, D. 1974. *Plants of Austin, Texas*—An Annotated Checklist of Native and Naturalized Plants. Saint Edward's University, 94 pp.

Sargent, C. S. 1933. *Manual of the Trees of North America*. Houghton Mifflin Co., Boston and New York, 910 pp.

Stanford, J. W. 1976. *Keys to the Vascular Plants of the Texas Edwards Plateau and Adjacent Areas*. Biology Department, Howard Payne University, Brownwood, Texas, 365 pp.

Vines, R.A. 1960. *Trees, Shrubs and Woody Vines of the Southwest*. The University of Texas Press. Austin, 1104 pp.

1977. *Trees of East Texas*. The University of Texas Press, Austin and London, 538 pp.

INDEX
of Common Names

INDEX
of Scientific Names